FIDIC YELLOW BOOK

The book analyses every aspect of the ease or otherwise of implementing the *FIDIC Yellow Book Conditions of Contract*. On a clause-by-clause basis, it highlights important structural features and suggests alternative text to avoid problems with the contract. Written in a user-friendly manner by an expert user of the FIDIC Suite of Contracts, who is a Member of the FIDIC President's List of Adjudicators, this book will be a vital reference point for contractors, lawyers, engineers, arbitrators and all others concerned with the FIDIC contracts.

Ben Beaumont is a active arbitrator, adjudicator and mediator as well as being a practicing Member of the Bar. He is a chartered surveyor, FRICS, FCIArb and a Chartered Arbitrator, as which he has more than 30 years of experience. He was a member of the original Marriott committee which was drawn together to draw up the draft Arbitration Law which is now the Arbitration Act 1996. He has adjudicated and arbitrated disputes arising from the FIDIC Suite of Contract Conditions. He is Founding Chair of the Forum for International Conciliation and Arbitration, www.ficacic.com.

CONTEMPORARY COMMERCIAL LAW

Causation in Insurance Contract Law
Meixian Song

Insurance Law in China
Edited by Johanna Hjalmarsson
and Dingjing Huang

Maritime Law in China: Emerging Issues and Future Developments
Edited by Johanna Hjalmarsson
and Jenny Jingbo Zhang

Illegality in Marine Insurance Law
Feng Wang

Insurance Law Implications of Delay in Maritime Transport
Ayşegül Buğra

Online Arbitration
Faye Fangfei Wang

Double Insurance and Contribution
Nisha Mohamed

The Law and Autonomous Vehicles
Matthew Channon, Lucy McCormick
and Kyriaki Noussia

FIDIC Yellow Book
A Commentary
Ben Beaumont

For more information about this series, please visit:
www.routledge.com/Contemporary-Commercial-Law/book-series/CCL

FIDIC YELLOW BOOK

A COMMENTARY

BEN BEAUMONT

informa law
from Routledge

First published 2019
by Informa Law from Routledge
2 Park Square, Milton Park, Abingdon, Oxon OX14 4RN

and by Informa Law from Routledge
52 Vanderbilt Avenue, New York, NY 10017

Informa Law from Routledge is an imprint of the Taylor & Francis Group, an informa business

British Library Cataloguing-in-Publication Data
A catalogue record for this book is available from the British Library

Library of Congress Cataloging-in-Publication Data
A catalog record for this book has been requested

ISBN: 978-1-138-05231-4 (pbk)
ISBN: 978-1-315-16786-2 (ebk)

Typeset in Times New Roman
by Apex CoVantage, LLC

TABLE OF CONTENTS

DETAILED CONTENTS

ACKNOWLEDGEMENTS

I acknowledge with great pleasure the support given to me by FIDIC in this enterprise, particularly Enrico Vink. I am also grateful to my publishing team Amy Jones and Caroline Church for patiently guiding me through the writing process. Finally, my heartfelt thanks go to Engineer Matthias Neuenschwander who has reviewed the text and spent days placing my script back upon the straight and narrow path.

GLOSSARY

There is contained in the FIDIC Contracts Guide a glossary of Contract terminology. It is somewhat surprising that this excellent glossary is not in all cases substituted for the definitions contained in these Conditions of Contract. In any event it is important to list this glossary at this juncture in this book.

This glossary lists words and phrases which are used in the fields of building, insolvency, Engineering and associated activities. The descriptions are not intended to amplify or replace the definitions in a FIDIC Contract, but are intended to indicate the overall principle involved. Initial letters are used to indicate words included elsewhere in the glossary.

Accepted Contract Amount. The amount which is offered by the Tenderer, and which is accepted by the Employer in the Letter of Acceptance, for providing Works in accordance with the Contract.

Addendum to Tender Documents. A document issued to each Tenderer during the Tender Period, amending the Tender Documents which were previously issued to each Tenderer.

Adjudication. A procedure under which a dispute is referred to an Adjudicator, which may include presentation of evidence, and which concludes with the Adjudicator's decision in accordance with the Contract. Generally, the decision will be binding, unless and until it is revised in an amicable settlement, arbitration or litigation.

Adjudicator. A person to whom a dispute between the Parties is referred for Adjudication.

Advance Payment. A payment (if specified in the Contract) which an Employer makes to a Contractor soon after the Effective Date, in order to finance the Contractor's mobilisation and other early expenditures.

Alternative Dispute Resolution. Resolution/settlement of a dispute using an alternative procedure to that which may be typically envisaged, as an alternative to arbitration or litigation.

Amicable Settlement. Resolution/settlement of the dispute by direct negotiation or by any other method which both Parties consider they can pursue amicably.

Appendix to Tender. The pages which are appended to the Letter of Tender, which defines various matters referred to in the Conditions of Contract.

Arbitration. A procedure under which a dispute has referred to one or 3 arbitrator(s), which includes a presentation of evidence, and which concludes with a binding award in accordance with the Contract. The procedure may be prescribed by law and/or by published rules of arbitration, and may be administered by an arbitral institution.

Arbitrator. A person to whom a dispute between the Parties is referred for arbitration.

Base Date. The date which is a specified period (usually 28 days) prior to the latest date for submission of the Tender to the Employer.

Bid and bidder. The Tender and the Tenderer: see below.

Bid-build. The procurement procedure under which the Employer provides most (or all) of the design details. The Contractor provides the Permanent Works in accordance with the Employer's design, and carries out the design of the Temporary Works and (possibly) particular parts of the Permanent Works as defined in the Contract.

Bill of Quantities. Under a Remeasurement Contract, the document which is issued by the Employer, and which is completed by the Tenderer and submitted with the Tender, as included in the Contract. This document contains, for each of the items of work, the physical quantity, unit rate and amount (or price) for the work. Each amount is a product of the quantity and the rate (in money per unit quantities).

Build-Operate-Train-Transfer (BOTT). The procurement procedure under which the concessionaire designs and provides completely equipped Works ready for operation, and then operates and maintains the Works during the Concession Period. The Works are then taken over by the project owner, whose operators will have previously been trained by the Concessionaire.

Build-Operate-Transfer (BOT). The procurement procedure under which the concessionaire designs and provides completely equipped Works ready for operation, and then operates and maintains the Works during the concession period. The Works are then taken over by the project owner.

Build-own-operate-transfer (BOOT). The procurement procedure under which the Concessionaire designs and provides completely equipped Works ready for operation, and then owns, operates and maintains the Works during the Concession Period. The Works then revert to the project owner.

Client. The Party for whom the Consultant carries out the Services under a Consultancy Agreement.

Commencement Date. The calendar date, of which the Contractor is given Notice, and which is the first day of the Time for Completion.

Concessionaire. Under BOT and similar procedures, the Party whose Tender has been accepted, and who is responsible for providing the Works and for operation and maintenance during the concession period; and the

Concession period. Under the BOT and similar procedures, the period during which the concessionaire operates and maintains the Works, and receives the monies which are typically intended cover the capital costs of providing the Works and also the costs of operation and maintenance. These monies may be based on the outcome of operation (for example, power generated or water treated), or other charges in respect of the use of, or end-product produced by, the facilities.

Conciliation. A procedure under which a dispute is referred to a conciliator, which may include presentation of evidence, and which may conclude with a binding agreement between the Parties. If there is no such agreement, the procedure is of no effect. The conciliator may then be required issue non-binding recommendations, which may take account of aspects other than those arising from the Contract.

Conciliator. The person to whom a dispute between the Parties is referred for conciliation.

Conditions of Contract. The document which specifies the fundamental stipulations of the Contract, the Parties' overall obligations under the Contract, and the allocation and consequences of risks.

Consortium. The joint venture comprising 2 or more Parties, each of which takes responsibility for a specific part of the obligations undertaken by the joint-venture.

Consultancy agreement. The services agreement under which the consultant carries out the services for the client, who may (for example) be the Employer or a design -build Contractor.

Consultant. A consulting Engineer or other professional, may be a person, an Engineering firm or a consortium (or other joint-venture) of such firms, and who carries out the services for a client under a consultancy agreement.

Contractor Contract Documents. The Documents which together constitute the legally enforcing all the agreement between the Parties, and which are usually named (and include) Conditions of Contract, Tender and Letter of Acceptance and/or Contract agreement. A Bid-build Contract will also include Employer's Requirements, Drawings, Bill of Quantities and (possibly) other Schedules. A design and build Contract will also include Employer's Requirements and proposal.

Contract Agreement. The formal agreement between the Parties, which confirms or establishes the Contract.

Contract Price. The amount, which is due and payable to the Contractor for providing the Works. Under a re-measurement Contract, The Contract Price is determined from the actual physical quantities of the completed work.

Contractor. The Party whose Tender has been accepted by the Employer, and who is responsible for providing the work; and the legal successors entitled to such Party.

Contractor's Documents. The calculations, computer programs and other software, Drawings, manuals, models and other Documents of a similar nature which the Contractor is required to provide, as specified in the Contract.

Contractor's Equipment. The machinery, apparatus and other things (other than temporary Works and Employer's Equipment, if any) which the Contractor requires in order to provide the Works. Contractor's equipment does not include anything, which is intended to form or forms part of the Permanent Works.

Contractor's Personnel. People whom the Contractor utilises on-Site, including: the Contractor's staff, labour and other Employers; the staff, labour and other Employees of each Subcontractor; and any other people appointed by the Contractor to assist in the execution of the Works.

Contractor's proposal. Under a design-build Contract, the proposal submitted with the Tender, is included in the Contract, which includes preliminary and/or outline design(s) of the Works.

Cost. The expenditure properly incurred (or to be incurred) by the Contractor, whether on or off the Site, including overhead and similar charges, but excluding Profit.

Country. The country in which the Works are to be provided and to which Plant and Materials are to be delivered.

Defects Notification Period. Period specified in the Contract for notifying defects, calculated from the date on which the Works (or, possibly, a Section) are completed and taken over.

Design Build. The procurement procedure under which the Contractor designs and provides the Works.

Design Build Finance Operate (DBFO). The procurement procedure under which the concessionaire obtains finance, designs and provides completely equipped Works ready for operation, and then operates and maintains the Works during the concession period, receiving payment in respect of the outcome of the operation (for example, power generated or water treated). The Works are then taken over by the project owner.

Dispute Adjudication Board (DAB). A panel of one or 3 persons, to whom a dispute between the Parties is initially referred for adjudication, which concludes with the board's decision in accordance with the Contract. Generally, the decision will be binding, unless and until it is revised in an amicable settlement, arbitration or litigation.

Dispute Review Board (DRB). A panel, usually of 3 persons to whom a dispute between the Parties is usually referred under a procedure which may include presentation of evidence and which generally concludes with the boards' non-binding recommendation for settlement of the dispute. The recommendation may become binding if neither Party gives the required Notice.

The Drawings. Under a Bid-Build Contract, the Drawings of the Works included in the Tender dossier, and any additional and modified Drawings issued by (or on behalf of) the Employer under the Contract.

Effective date. The calendar date on which the Contract comes into full force and effect.

Employer. The Party who receives and accepts the Tender, who is responsible for providing the Site and paying the Contractor, and for whom the Works are provided by the Contractor; and the legal successors entitled to such a Party.

Employer's Equipment. The machinery and apparatus (if any) which the Employer is required to make available for the use of the Contractor, as specified in the Contract.

Employer's Personnel. People whom the Employer utilises on-Site (other than Contractor's Personnel), may include: the Employer's staff, Labour and other Employees; and the staff, labour and other Employees of the Engineer or Employer's Representative.

Employer's Requirements. Under a bid-build Contract, the technical document issued by the Employer to the Tenderer, as included in the Contract and any additions and modifications to such Employer's Requirements in accordance with the Contract. Whereas the Employer's Requirements usually specifies all the technical Requirements for the Works.

Employer's Requirements. Under a design-build Contract, the document issued by the Employer to the Tenderer, as included in the Contract, and any additions and modifications to such Requirements in accordance with the Contract. This document may specify the required skill, standard, design criteria and other Requirements for the Works.

Engineer. A Consulting Engineer or other professional, whom the Contract requires the Employer to appoint, and who may be the person, an Engineering firm or a consortium (or other joint-venture) of such firms.

Engineer, Procure and Construct (EPC) Contract. The procurement procedure under which the Contractor or supplier designs and provides Works which are fully complete and ready for operation by the Employer. An EPC Contract may also require the Contractor to provide finance for the capital costs of providing the Works, and/or consumables for a specified period of operation and maintenance.

FICA. An UNCITRAL Observer NGO with special interests in dispute avoidance www.ficacic.com

Final statement. The statement submitted by the Contractor in order to receive final payment, as agreed.

Force majeure. An Exceptional Event or circumstance which satisfies the criteria for force majeure stated in the Contract.

Foreign Currency. The Currency, specified in the Contract as a Currency in which the Contractor is due to be paid, other than the local Currency.

Goods. Contractor's equipment, Materials, Plant and temporary Works, or any part of them as appropriate.

ICC. The International Chamber of Commerce, which is based at 38 Cours Albert 1er, 75008, Paris, France. See www.iccwbo.org.

Instructions to Tenderers. The document which is issued to the Tenderer with the Tender dossier, which contains details applicable the Tender period (for sample, the arrangements for visiting the Site and for receipt and opening of Tenders), and which should therefore not become part of the Contract.

Joint venture. 2 or more Parties who, together, undertake a venture for profit.

Letter of Acceptance. The Employer's letter to the Tenderer, formally accepting the offer contained in the Tender.

Letter of invitation. The Employer's letter to the Tenderer, or the published advertisement(s), containing the invitation to prepare and submit a Tender.

Letter of Tender. The Tenderer's letter to the Employer, formally offering to provide the Works.

Local Currency. The Currency of the country in which the Works are to be provided and to which Plant and Materials are to be delivered.

Lump Sum Contract. The form of Contract under which the Tenderer offers lump sum prices for providing the Works, thereby taking the risk that the physical quantities of the Works may differ from those which he had calculated in order to finalise his Tender sum.

Materials. Things of all kinds (other than Plant) which the Contractor is required to supply and incorporate into the Permanent Works, as specified in the Contract.

Mediation. A procedure under which a dispute is referred to mediator, which may include presentation of evidence, and which may conclude with a binding agreement between the Parties. If there is no such binding agreement, the procedure is of no effect. The mediator may then be required to issue non-binding recommendations, which may take account of aspects other than those arising from the Contract.

Nominated Subcontractor. A Subcontractor whom the Contractor is required (by the Employer or by the terms of the Contract) to appoint, but for whom the Contractor retains responsibility; and the legal successors entitled to such Subcontractor.

Open Tendering. The procurement procedure under which any person may obtain a Tender dossier and submit a Tender for the Works, without prequalification. However, post-qualification may be required.

Party. One of the 2 Parties to enter into a Contract with the other, and who are typically the Employer and the Contractor.

Payment certificate. A certificate which is issued under the Contract, and which states the amount which the certifier considers that the Contractor is then due to be paid.

Performance Certificate. The certificate which is issued under the Contract when the specified certifier considers that the Contractor has performed all obligations under the Contract.

Performance Security. The security for proper performance (usually, bond or bank guarantee), which the Contractor is required to provide, as specified in the Contract. The Performance Security provides protection for the Employer in case of major default by the Contractor.

Permanent Works. The Works which the Contractor is required to provide and hand over to the Employer, as specified in the Contract.

Plant. The machinery and apparatus which the Contractor is required to provide and hand over to the Employer, as specified in the Contract.

Post-qualification. The part of an open Tendering procedure, which is carried out after receipt of Tenders, and during which the Employer assesses Tenderer's ability the form of Contract.

Privately financed infrastructure project. The project for which the concessionaire obtains finance, designs and provides completely equipped Works ready for operation, then operates and maintains the Works during the concession period, receiving payment in respect of the outcome of operation (for example, power generated or water treated). The Works are then taken over by the project owner.

Privatisation. The process of converting a public utility into private ownership.

Programme. Codified instructions used by a computer; or the USA spelling for a programme.

Programme. A document which is typically prepared and issued by the Contractor, and which shows the proposed timing of each stage of the provision of the Works.

Proposal. For a consultancy agreement: the consultants offer to carry out the services. For a design and build Contract: the Tenderer's proposal to provide the Works, which becomes the Contractor's proposal.

Provisional sum. A sum specifically described in the Contract as a Provisional Sum, for the execution of any part of the Works all the supply of Plant, Materials or services. Being provisional, the payment due to the Contractor will depend on the work which he is required to carry out, if any.

Re-measurement Contract. The form of Contract under which the Tenderer prices a Bill of Quantities, and under which the payment due to the Contractor is based on the actual physical quantities of the completed Works.

Representative. The person representing the relevant Party defined in the Contract.

Repair Operate Transfer (ROT). The procurement procedure under which the concessionaire repairs and refurbishes an existing facility ready for operation, the then operates and maintains the facility during the concession period. The facility is then handed back to the project owner.

Retention money. The accumulated monies (if specified in the Contract) which are deducted and retained by the Employer from the payments otherwise due to the Contractor, which are only paid after completion.

Schedules. The document which is issued by the Employer, and which is completed by the Tenderer and submitted with the Tender, as included in the Contract. This document contains data and information which the Employer wishes the Tenderer to specify and which the Parties wish would become parts of the Contract.

Section. A part of the Works, which is specifically defined in the Contract is a Section, and which may be required to be completed by the Contractor and taken over by the Employer before other parts of the Works.

Services. The services which a consultant is required to provide, as stated in its consultancy agreement.

Site. The places provided by the Employer, whether Works are to be executed and to which Plant and Materials are to be delivered.

Statement. Statement submitted by the Contractor as part of an application for payment.

Subcontractor. The Party whom the Contractor appoints to provide a part of the Works, for whom the Contractor retains responsibility under the Contract, and the legal successors entitled to such Party.

Taking-Over Certificate. A Certificate which is issued under the Contract, when the specified certifier considers that the whole (or part) of the Works has been completed and is ready for the Employer to take over.

Temporary Works. All temporary Works of every kind (other than Contractor's equipment) required for the execution and completion of the Works and the remedying any defects.

Tender. The set of Documents which the Tenderer has completed and submitted to the Employer.

Tender Documents. The set of Documents which: the Employer issues to the Tenderer (and which is sometimes referred to as the Tender dossier), or the Tenderer submits to the Employer.

Tender dossier. The set of Documents which the Employer issues to a Tenderer for him to prepare and submit a Tender.

Tender Agent. The person identified by the Employer who receives the Tender dossier and subsequently submits the Tender for acceptance by the Employer.

Tender period. The period which commences when each Tenderer can obtain a Tender dossier and which expires on the latest day for the submission of the Tender for acceptance by the Employer.

Tender sum. The sum offered by the Tenderer for providing the Works in accordance with the Contract.

Tender validity. The calendar date stated in the Tender up to which it is open for acceptance by the Employer; or the period which expires on such date, calculated from the latest date for submission of the Tender.

Terms of reference. That document which is typically prepared by the client and sent to a prospective consultant, and which describes the proposed scope of the services.

Tests after Completion. Tests which are required to be carried out after the Works (or, possibly, a Section) are complete and taken over by the Employer, all in accordance with the Contract.

Tests on Completion. Tests which the Contractor is required to carry out before the Works (or, possibly, Section) are taken over by the Employer, all in accordance with the Contract.

Time for Completion. The period specified in the Contract as that within which the Contractor is required to complete the Works (or, possibly, a Section), calculated from the commencement date.

Turnkey. The procurement procedure under which the Contractor or supplier designs and revised Works, which are ready for operation by the Employer (at the "turn of the key"), and which may incorporate all the necessary fixtures, fittings and equipment.

UNCITRAL. United Nations Commission on International trade Law. See www.uncitral.org

Value Engineering. A procedure under which an Engineering design is reviewed with the intention of improving the value to the Employer of the completed Works.

Variation. An alteration or modification to such Works as a Contractor is required to provide under the Contract. Under a bid-build Contract, a variation may be ordered by a change to the Employer's Requirements or drawing. Under a design-build Contract, the variation may be ordered, or maybe approved after being proposed by the Contractor's designers.

Works. The Permanent Works and the temporary Works, which the Contract requires the Contractor to provide.

World Bank. The International bank for Reconstruction and development and the International Development Association, which are based at 1818H Street NW, Washington DC 20433, USA. See www.worldbank.org

CHAPTER 1

Clause 1 General Provisions

1.1 Definitions

1.1 In the Contract following words and expressions shall have the meaning stated, except where the context requires otherwise.

This is a fundamentally unhelpful comment as an introduction to the general Conditions. The opportunity to identify a context, which requires a different interpretation from that set out in the general definitions is impossible to define.

1.1.1 "**Accepted Contract Amount**" means the amount accepted in the Letter of Acceptance for the execution of the Works in accordance with the Contract.

This definition is accurate providing that there were no Conditions imposed in the Letter of Acceptance, which could or would vary the Accepted Contract Amount should those Conditions take effect.

1.1.2 "Advance Payment Certificate" means a Payment Certificate issued by the Engineer for advance payment under Sub-Clause 14.2.2 [*Advance Payment Certificate*].

There are various steps, which must be complied with before the certificate can be issued and these are set out in Sub-Clause 14.2.2 [*Advance Payment Certificate*].

1.1.3 "**Advance Payment Guarantee**" means the guarantee under Sub-Clause 14.2.1 [*Advance Payment Guarantee*].

This guarantee is intrinsically linked to the procedure enabling the Contractor to receive payment under the Advance Payment Certificate structure.

1.1.4 "**Base Date**" means the date 28 days before the latest date for the submission of the Tender.

It is important to note that it is implied that the Base Date can vary should the Employer decide to extend either once or more than once the date for the submission of the Tender. The precise Determination of the Base Date can be critical when taking into account regulations or legislation which come into force just after the Base Date and which have therefore to be taken into account under the Contractor's obligations ensure that the Design and Execution of the Works and/or Plant are fully compliant with all the Laws and regulations of the Country. On occasions compliance with new legislation will engender variations which will incur a cost to the Employer.

1.1.5 "**Claim**" means a request or assertion by one Party to the other Party for an entitlement or relief under any Clause of these Conditions or otherwise in connection with, or arising out of, the Contract or the execution of the Works.

The use of the phrase an entitlement suggests a reward different from money. The definition of Claim might be better phrased as follows:

> *"'Claim' means a request by one Party to the other Party for a monetary payment and/or for an extension of time, where appropriate, or other relief of whatever nature under any Clause of these Conditions or otherwise in connection with, or arising out of, the Contract or the execution and operation of the Works and/or Plant."*

The amended definition allows for an explicit Claim to arise from the operation of the Works or Plant and further specifically identifies that an area of Claim may arise out of the operation of the Plant. That issue is not clear from the original definition.

1.1.6 "**Commencement Date**" means the date as stated in the Engineer's Notice issued under Sub-Clause 8.1 [*Commencement of Works*].

There may be occasions where a Letter of Acceptance has not been provided but the Engineer has issued a Notice under Sub-Clause 8.1. In that event the Contractor may safely proceed to commence operation under his obligations under the Contract.

1.1.7 "**Compliance Verification System**" means the compliance verification system to be prepared and implemented by the Contractor for the Works in accordance with Sub-Clause 4.9.2 [*Compliance Verification System*].

This requirement is now usual for all contracts.

1.1.8 "**Conditions of Contract**" or "these Conditions" means these General Conditions as amended by the Particular Conditions.

This is a very unhelpful definition. What is intended is as follows:

> *"Conditions of Contract or these Conditions mean the general Conditions of Contract and such particular Conditions as may form part of the Contract. These particular Conditions may add to or vary the general Conditions."*

1.1.9 "**Contract**" means the Contract Agreement, the Letter of Acceptance, the Letter of Tender, any addenda referred to in the Contract Agreement, these Conditions, the Employer's Requirements, the Schedules, the Contractor's Proposal, the JV Undertaking (if applicable) and the further documents (if any), which are listed in the Contract Agreement or in the Letter of Acceptance.

This definition is too restrictive. The definition should be phrased as follows:

> *"'Contract' means any combination being: the Contract agreement, the Letter of Acceptance, a Letter of Tender, any additional Documents referred to in the Contract agreement, these Conditions, the Employer's Requirements, the Schedules, the Contractor's proposal, the Particular Conditions, any JV Undertaking (if applicable) and any further Documents (if any) which are listed in the Contract Agreement and/or in the Letter of Acceptance."*

This definition is more loosely drafted than that provided in the definition stated above and gives both Parties greater flexibility. It is very important that Documents can be read in the alternative in order to provide an effective binding legal context.

1.1.10 "**Contract Agreement**" means the agreement entered into by both Parties in accordance with Sub-Clause 1.6 [*Contract Agreement*].

As Sub-Clause 1.6 anticipates it is not essential for the Parties to enter into a Contract agreement. The key document is the Letter of Acceptance.

1.1.11 "**Contract Data**" means the pages, entitled contract data, which constitute Part A of the Particular Conditions.

There is no reason for the information which constitutes Contract Data should be placed in a document entitled to Contract Data. Further "Contract Data" does not mean the pages, it refers to the information provided by the Employer to the Contractor and upon which the Contractor may well base the Contract amount. Therefore, the importance placed upon carefully produced and detailed Contract Data cannot be underestimated.

A better definition would be:

> *"Contract Data means that the relevant information provided by the Employer and where appropriate the Contractor in written form whether entitled Contract Data or not upon which the Contractor relies when determining the bid offer."*

1.1.12 "**Contract Price**" means the price defined in Sub-Clause 14.1 [*The Contract Price*].

This is self-explanatory.

1.1.13 "**Contractor**" means the person(s) named as contractor in the Letter of Tender accepted by the Employer and the legal successors in title of such person(s).

This is self-explanatory.

1.1.14 "**Contractor's Documents**" means the documents prepared by the Contractor as described in Sub-Clause 5.2 [*Contractor's Documents*], including calculations, digital files, computer programs and other software, drawings, manuals, models, specifications and other documents of a technical nature.

Sub-Clause 5.2 does in fact not provide any detail of a definition of Contractor's Documents. The only question here is raised over the use of the word "supply". The word "supplying" implies a giving by someone to someone, i.e. the Employer or where appropriate the Engineer. Questions are raised as to the situation at the moment of the supply of the documents (during Tender or during realisation of the Works), and what happens to these documents when approved or following a non-objection notice: are they still documents upon which the Contactor relies (only), or are may they become binding for the Employer also?

In fact, the items listed under this definition while fundamental to the construction of the Permanent Works are not necessarily going to form part of the Permanent Works. An alternative phrase might be more appropriate "prepared by the Contractor for his use under the Contract."

This definition should be loosened in order to allow for most possibilities as follows:

> *"'Contractor's Documents' means the Documents to be supplied by the Contractor as described in Sub-Clause 5.2 Contractor's Documents, and which may include: calculations, digital files, Computer programs and other software, drawings, provisional and final operational manuals, models, specifications and all other Documents whether of a technical nature or otherwise upon which the Contractor relies in order to fulfil his obligations under the Contract."*

1.1.15 "**Contractor's Equipment**" means all apparatus, equipment, machinery, construction plant, vehicles and other items required by the Contractor for the execution of the Works. Contractor's Equipment excludes Temporary Works, Plant, Materials and any other things intended to form or forming part of the Permanent Works.

This definition has excluded any reference to remedying of the defects. Previously Contractor's equipment was envisaged being used during the defects notification period. It is not clear why this period has been impliedly excluded.

1.1.16 "**Contractor's Personnel**" means the Contractor's Representative and all personnel whom the Contractor utilises on Site or other places where the Works are being carried out, including the staff, labour and other employees of the Contractor and of each Subcontractor; and any other personnel assisting the Contractor in the execution of the Works.

The Reference to any other Personnel assisting the Contractor in the execution of the Works shall include any independent personnel who assist the Contractor before, during and/or after the execution of the Works.

1.1.17 "**Contractor's Proposal**" means the part of the Tender stated or implied as being the Contractor's proposal for execution of the Works, as included in the Contract. Such documents may include the Contractor's preliminary design.

This definition where stating a reference to an Implied Proposal which takes account of the fact that on occasions Documents which should be entitled or headed in a certain manner are not so headed or do not yet include the key information which enable them to form part of the Contract and/or the Contractor's Documents.

1.1.18 "**Contractor's Representative**" means the natural person named by the Contractor in the Contract or appointed by the Contractor under Sub-Clause 4.3 [*Contractor's Representative*], who acts on behalf of the Contractor.

This Definition is not quite clear. A redraft would be:

> *"'Contractor's Representative' means the natural person named by the Contractor in the Contract or appointed by the Contractor under Sub-Clause 4.3 Contractor's Representative, as the Contractor's Representative and who acts on behalf of the Contractor with respect to all matters relevant to the progress of the Works."*

1.1.19 "**Cost**" means all expenditure reasonably incurred (or to be incurred) by the Contractor in performing the Contract, whether on or off the Site, including taxes, overheads and similar charges, but does not include profit. Where the Contractor is entitled under a Sub-Clause of these Conditions to payment of Cost, it shall be added to the Contract Price.

A more effective definition is as follows:

> *"'Cost' means or expenditure reasonably incurred and including any charge for Daywork (or to be incurred) by the Contractor performing his obligations under the Contract, whether on or off the Site, including material and labour cost, taxes, overheads and similar charges but does not include profit. Where the Contractor is entitled under a Sub-Clause of these Conditions to payment of Cost, it shall be added to the Contract Price."*

1.1.20 "**Cost Plus Profit**" means Cost plus the applicable percentage for profit stated in the Contract Data (if not stated, five percent (5%)). Such percentage shall only be added to Cost, and Cost Plus Profit shall only be added to the Contract Price, where the Contractor is entitled under a Sub-Clause of these Conditions to payment of Cost Plus Profit.

This is an unwieldy definition.

First sentence is clear. Thus a revised definition would be:

> *"'Cost Plus Profit' means costs plus the applicable percentage profit stated in the Contract Data (if not stated, then the percentage will be five percentum (5%)). Such percentage shall only be added to cost where the Contractor is entitled under a Sub-Clause of these Conditions to payment of Cost Plus Profit.*
>
> *The calculation being Cost Plus Profit shall thereafter be added to the Contract Price."*

1.1.21 "**Country**" means the country in which the Site (or most of it) is located, where the Permanent Works are to be executed.

This is not accurate. What is intended is that country where the Works and/or Plant are located is to be identified. Therefore a better definition is:

> *"Country means the country where the Permanent Works and/or Plant is located."*

1.1.22 "**DAAB**" or "**Dispute Avoidance/Adjudication Board**" means the sole member or three members (as the case may be) so named in the Contract, or appointed under Sub-Clause 21.1 [*Constitution of the DAAB*] or Sub-Clause 21.2 [*Failure to Appoint DAAB Member(s)*].

1.1.23 "**DAAB Agreement**" means the agreement signed or deemed to have been signed by both Parties and the sole member or each of the three members (as the case may be) of the DAAB in accordance with Sub-Clause 21.1 [*Constitution of the DAAB*] or Sub-Clause 21.2 [*Failure to Appoint DAAB Member(s)*], incorporating by reference the General Conditions of Dispute Avoidance/Adjudication Agreement contained in the Appendix to these General Conditions with such amendments as are agreed.

This definition could be simplified.

> *"DAAB agreement means the agreement entered into by both Parties on the one hand and the sole member or each of the three members (as the case may be) of the DAAB in accordance with Sub-Clause 21.1[Constitution of the DAAB]. The agreement may incorporate the terms of the dispute avoidance/ adjudication board (DAAB) rules published by FIDIC (or any subsequent published amendment or modification of these rules being in force at the Base Date) together with such amendments to the agreement as are agreed on the one hand by both Parties and on the other by members of the DAAB."*

1.1.24 "**Date of Completion**" means the date stated in the Taking-Over Certificate issued by the Engineer; or, if the last paragraph of Sub-Clause 10.1 [*Taking Over the Works and Sections*] applies, the date on which the Works or Section are deemed to have been completed in accordance with the Contract; or, if Sub-Clause 10.2 [*Taking Over Parts*] or Sub-Clause 10.3. [*Interference with Tests on Completion*] applies, the date on which the Works or Section or Part are deemed to have been taken over by the Employer.

1.1.25 Day means a calendar day

1.1.26 "**Daywork Schedule**" means the document entitled daywork schedule (if any) included in the Contract, showing the amounts and manner of payments to be made to the Contractor for labour, materials and equipment used for daywork under Sub-Clause 13.5 [*Daywork*].

A more all-encompassing definition might be:

> *"Daywork schedule means the document included in the Contract or agreed between the Parties showing the rates applicable to each item of labour, Materials and equipment used by the Contractor for daywork under Sub-Clause 13.5 [Daywork]. Where an item of labour, Materials or equipment is absent from the schedule then the Engineer shall determine a fair rate for the item"*

1.1.27 "**Defects Notification Period**" or "**DNP**" means the period for notifying defects and/or damage in the Works or a Section or a Part (as the case may be) under Sub-Clause 11.1 [*Completion of Outstanding Work and Remedying Defects*], as stated in the Contract Data (if not stated, one year), and as may be extended under Sub-Clause 11.3 [*Extension of Defects Notification Period*]. This period is calculated from the Date of Completion of the Works or Section or Part.

The last sentence of this definition needs to be re-written.

> *"This period is calculated from the date on which the Works or Section is taken over or deemed to be taken over under Sub-Clause 10.1 [Taking Over the Works and Sections] or a Part is taken over or deemed to be taken over under Sub-Clause 10.2[Taking over Parts.]"*

The deeming provision applies whether the taking over is of the Works, Sections or Parts. Therefore the definition should reflect that fact.

1.1.28 "**Delay Damages**" means the damages for which the Contractor shall be liable under Sub-Clause 8.8 [*Delay Damages*] for failure to comply with Sub-Clause 8.2 [*Time for Completion*].

This definition is self-explanatory.

1.1.29 "**Dispute**" means any situation where:

(a) one Party makes a claim against the other Party (which may be a Claim, as defined in these Conditions, or a matter to be determined by the Engineer under these Conditions, or otherwise);
(b) the other Party (or the Engineer under Sub-Clause 3.7.2 [*Engineer's Determination*]) rejects the claim in whole or in part; and
(c) the first Party does not acquiesce (by giving a NOD under Sub-Clause 3.7.5 [*Dissatisfaction with Engineer's determination*] or otherwise), provided however that a failure by the other Party (or the Engineer) to oppose or respond to the claim, in whole or in part, may constitute a rejection if, in the circumstances, the DAAB or the arbitrator(s), as the case may be, deem it reasonable for it to do so.

This is an extraordinarily complex and unusual definition.

In the first leg of the definition the explanatory Clause is placed after the word Party instead of after the word "Claim". Secondly, most participants involved in these Conditions of Contract know what a Claim is. It is a request by Party A for something from Party B. If there is a concession from Party B that the request is justified there is no dispute. However, if the request is rejected wholly or in part then there is a dispute as to the element rejected.

What this definition has done is to create a scenario where by Party B may, by silence, indicate its rejection of the Claim. Without citing every instance there are time limits, which apply within these Conditions of Contract as a result of which deeming provision come into force if the responding Party is silent. The situation described in this definition reverses that concept. As a result, a Contractor, and it will usually be the Contractor, will be in the position of complete uncertainty as to whether or not a Claim has been rejected. This is a most extraordinary proposition.

Therefore, not only is this definition over complex but it has turned the Claim process into a situation where an Employer, or Engineer, may by silence, which is reasonable, reject the Claim. How can silence be reasonable?

The tried and tested formula for describing a dispute should be substituted for this extraordinary conglomeration of words. Also, the addition of a claim that is not a Claim to the concept is strange and should be avoided. Therefore:

> *"Dispute means:*
> *Where one Party makes a Claim against the other Party;*
> *as a result the other Party rejects the Claim wholly or in part in a timely matter as set out in these Conditions of Contract, where appropriate; and*
> *the Claiming Party does not accept the rejection as bringing an end to the matter."*

1.1.30 **"Employer"** means the person named as the Employer in the Contract Data and the legal successors in title to this person.

This definition should be widened to include:

> *"the person named as the Employer in the Contract Data or/and on the Letter of Acceptance or whichever other document rates the highest in the priority list of documents and the legal successors in title to this person."*

It may well be that person, Natural or otherwise, named in the Contract Data is not the Party who is the Employer and therefore it is appropriate to look to the person identified on the Letter of Acceptance or whichever other document rates the highest in the priority list of documents as deciding the matter as to who is the Employer.

1.1.31 **"Employer's Equipment"** means the apparatus, equipment, machinery, construction plant and/or vehicles (if any) to be made available by the Employer for the use of the Contractor under Sub-Clause 2.6 [*Employer-Supplied Materials and Employer's Equipment*]; but does not include Plant which has not been taken over under Clause 10 [*Employer's Taking Over*].

It is not clear how the exclusion at the end of the definition assists the definition. Employer's Equipment appears to refer to all items loaned or leased by the Employer to the Contractor solely and simply for the purpose of executing the Works. By its own definition it could not include Plant, which has been taken over by the Employer and forms part of the Permanent Works. But it certainly could include Plant, which belongs to the Employer and is either apparatus machinery or vehicles. There is a difference between construction plant and Plant, which is a defined term. Plant made available to the Contractor by the Employer is closer to Material made available by the Employer to the Contractor than Equipment, as it will form part of the Works. The definition is not very elegant but it is correct.

1.1.32 "**Employer's Personnel**" means the Engineer, the Engineer's Representative (if appointed), the assistants described in Sub-Clause 3.4 [*Delegation by the Engineer*] and all other staff, labour and other employees of the Engineer and of the Employer engaged in fulfilling the Employer's obligations under the Contract; and any other personnel identified as Employer's Personnel, by a Notice from the Employer or the Engineer to the Contractor.

A slight comment to this definition would be to remove the comma after Employer's Personnel in the last line. The Notice requirement identified does not need to be separated from the various Personnel identified as Employer's Personnel but needs to be linked to it without separation by a comma.

1.1.33 "**Employer's Requirements**" means the document entitled Employer's Requirements, as included in the Contract, and any additions and modifications to such document in accordance with the Contract. Such document describes the purpose(s) for which the Works are intended, and specifies Key Personnel (if any), the scope, and/or design and/or other performance, technical and evaluation criteria, for the Works.

Again, this definition is too restrictive.

The key definition should be as follows:

> *"Employer's Requirements mean the document or Documents which describes the purposes and scope for which the Works are intended and the product (if any) intended to be produced and, where appropriate, specifies the design and, where appropriate, other performance, technical and evaluation criteria applicable to the Works."*

While there is no need for the document to be entitled or headed Employer's Requirements, it would be preferable that the Employer's Requirements (be wrapped up in a document called "Employer's Requirements": this facilitates the Contractor's task and obliges the Employer to be a little orderly in his Tender docs. A special paragraph about this should go into the Guidance on Preparation of Tender Documents. What is vital is that there is a document, which describes exactly what the Employer requires and how the Employer envisages that product being created.

1.1.34 **Employer-Supplied Materials**" means the materials (if any) to be supplied by the Employer to the Contractor under Sub-Clause 2.6 [*Employer-Supplied Materials and Employer's Equipment*].

This definition is linked to 1.1.31 [*Employer's Equipment*]. That definition encompasses the equipment whereas this definition encompasses the Materials.

1.1.35 "**Engineer**" means the person named in the Contract Data appointed by the Employer to act as the Engineer for the purposes of the Contract, or any replacement appointed under Sub-Clause 3.6 [*Replacement of the Engineer*].

Again, this Clause is self-explanatory save that the definition does not allow for the Employer to appoint the Engineer after the Contract Data has been compiled. It would be preferable to state:

> *"Engineer means the person named in the Contract Data and appointed by the Employer to act as the Engineer for the purposes of the Contract or any replacement appointed on the Sub-Clause 3.6 [Replacement of the Engineer]."*

The comma after the word "Contract" in the second line has been removed in the amended definition as the replacement option is meant to be read together with the first part of the definition.

1.1.36 "**Engineer's Representative**" means the natural person who may be appointed by the Engineer under Sub-Clause 3.3 [*Engineer's Representative*].

The provisions for appointment are set out in Sub-Clause 3.3.

1.1.37 "**Exceptional Event**" means an event or circumstance as defined in Sub-Clause 18.1 [*Exceptional Events*].

It is difficult to determine what the addition of the words "or circumstance" adds to the definition. "Event" according to the Merriam-Webster web dictionary is "something that happens" (occurrence), while "circumstance would be "a condition, fact, or event accompanying, conditioning, or determining another : an essential or inevitable concomitant *the weather is a circumstance to be taken into consideration"*. If circumstance is complementary to event, then the meaning would include condition and fact. The authors may have meant "condition": the contract regards the conditions (physical and other) under which the Works are realised. Maybe "condition" would be a better word.

1.1.38 "**Extension of Time**" or "**EOT**" means an extension of the Time for Completion under Sub-Clause 8.5 [*Extension of Time for Completion*].

This is the first time that the concept of an extension of time has been formally identified in the definition section of these Conditions.

1.1.39 "**FIDIC**" means the Fédération Internationale des Ingénieurs-Conseils, the International Federation of Consulting Engineers.

1.1.40 "**Final Payment Certificate**" or "**FPC**" means the payment certificate issued by the Engineer under Sub-Clause 14.13 [*Issue of FPC*].

This definition merely refers to the detail of the Sub-Clause, which addresses procedure by which the Engineer is put in the position to issue the Final Payment Certificate.

1.1.41 "**Final Statement**" means the Statement defined in Sub-Clause 14.11.2 [*Agreed Final Statement*].

It is seriously considered that there should be a separate definition for Agreed Final Statement. However, Sub-Clause 14.11.2 appears to spend the majority of the text examining the situation where an agreed final statement will not be forthcoming. In general, it would be preferable to include all definitions in this section, so as to avoid the necessity of definitions elsewhere. It seems strange that a definition in the definition section refers to a definition elsewhere.

1.1.42 "**Foreign Currency**" means a currency in which part (or all) of the Contract Price is payable, but not the Local Currency.

It is not clear why there is a need for a reference to the Currency being foreign. It would be sufficient to refer to the Currency as being that in which part or all of The Contract Price is payable but excluding the local Currency. Further it would be better to refer to Currency as the Currencies, as opposed to any Currency, in which part or all of the Contract Price is payable. It is a common occurrence for the Agreed Contract Amount or and the Contract Price to be composed of a basket of Currencies. The definition as currently drafted raises issues as it refers to a singular Currency instead of all possible Currencies.

It is not a clear definition. A better definition would be:

> *"Foreign Currency means any Currency other than the local Currency in which part (or) of the Contract price is payable."*

1.1.43 "**General Conditions**" means this document entitled "Conditions of Contract for Plant and Design-Build for Electrical & Mechanical Plant, and for Building and Engineering Works, designed by the Contractor", as published by FIDIC.

It is not absolutely clear why the definition at 1.1.8 Conditions of Contract for these Conditions is so very separate as to only represent General Conditions as described in the previous definition which are amended by the particular Conditions.

1.1.44 "**Goods**" means Contractor's Equipment, Materials, Plant and Temporary Works, or any of them as appropriate.

Contractor's Equipment is defined earlier in the general Conditions of Contract. It is suggested that to avoid all doubt this definition should be amended as follows:

> *"Contractor's Goods means Contractor's Equipment, Materials, Plant and such Temporary Works as are necessary for the execution of the Works or any of them as are appropriate."*

1.1.45 "**Interim Payment Certificate**" or "**IPC**" means a Payment Certificate issued by the Engineer for an interim payment under Sub-Clause 14.6 [*Issue of IPC*].

This definition is self-explanatory and the comment made under 1.1.42 applies.

1.1.46 "**Joint Venture**" or "**JV**" means a joint venture, association, consortium or other unincorporated grouping of two or more persons, whether in the form of a partnership or otherwise.

Here are general Conditions of Contract that are attempting to impose a generic term of art being joint-venture, upon a variety of authorised and legal corporate combinations which may not in fact be a joint-venture.

1.1.47 "**JV Undertaking**" means the letter provided to the Employer as part of the Tender setting out the legal undertaking between the two or more persons constituting the Contractor as a JV. This letter shall be signed by all the persons who are members of the JV, shall be addressed to the Employer and shall include:

(a) each such member's undertaking to be jointly and severally liable to the Employer for the performance of the Contractor's obligations under the Contract;
(b) identification and authorisation of the leader of the JV; and
(c) identification of the separate scope or part of the Works (if any) to be carried out by each member of the JV.

These definitions do not require that the JV, were it to exist, to disclose to the Employer the terms of the JV. Person is not defined in these general Conditions of Contract. It is clear that these Conditions anticipate that a person may be a natural person or a body corporate. It would be more appropriate in this definition to amend the second sentence as follows:

> *"This letter shall be signed by all the Representatives of the members of the JV and shall be addressed to the Employer..."*

It is also a harsh reality in these construction businesses that in a joint-venture the Contract imposes joint and several liabilities upon all the joint-venture partners however minimal is the contribution of the particular partner.

The requirement for each member to be jointly and severally liable for this to be carefully considered by any Contractor forming part of the joint-venture. The reason is simple. Not all joint-venture partners are equal partners. And the junior partners need to ensure that they are effectively and properly protected by insurance and other contractual arrangements against wilful or other acts by the predominant member or members of the joint-venture.

The next requirement is that the Undertaking must identify the leader and rather baldly addresses authorisation of the leader of the JV.

The leader's role may be practical or purely administrative. Some model Contracts for joint-ventures anticipate that the leader shall be no more than a mouthpiece and shall have virtually no power while other model Contracts give the leader powers to bind other members of the consortium or joint-venture.

It would be better for the Employer to state his requirements regarding authority of the leader in his Tender Documents.

Finally, this definition requires that the Undertaking identifies the separate scope of work or part of the work to be carried out by each member of the *Joint-Venture*. That is a rational requirement and is a very important document for future negotiation. As this identification of the different areas of operation of the *Joint-Venture* Parties is key to a smooth running and effective Contract management system.

1.1.48 "**Key Personnel**" means the positions (if any) of the Contractor's Personnel, other than the Contractor's Representative, that are specified in the Employer's Requirements.

1.1.49 "**Laws**" means all national (or state or provincial) legislation, statutes, acts, decrees, rules, ordinances, orders, treaties, international law and other laws, and regulations and by-laws of any legally constituted public authority.

The only issue with this definition is its excessive search for different words meaning laws. It might be preferable to define laws as:

> *"Laws means national, state or provincial legislation or regulation of whatever nature published by a legally constituted public authority."*

1.1.50 "**Letter of Acceptance**" means the letter of formal acceptance, signed by the Employer, of the Letter of Tender, including any annexed memoranda comprising agreements between and signed by both Parties. If there is no such Letter of Acceptance, the expression "Letter of Acceptance" means the Contract Agreement and the date of issuing or receiving the Letter of Acceptance means the date of signing the Contract Agreement.

It is very important to give flexibility to the meaning of Letter of Acceptance. Frequently a Contract will be commenced on the basis of the sending of an email. There may not be anything more formal than a responding email of acceptance. The definition of Contract agreement defines such an agreement as the agreement entered into by both Parties in accordance with Sub-Clause 1.6 [*Contract Agreement*]. Minimal legal Requirements are necessary in order to constitute such an agreement.

The last sentence of the definition is unclear. A simpler definition would be:

> *"If there is no such Letter of Acceptance, the expression Letter of Acceptance means the document constituting the Contract Agreement and the date of issuing the Letter of Acceptance means the date when the Employer signed the Contract Agreement."*

1.1.51 "**Letter of Tender**" means the letter of tender, signed by the Contractor, stating the Contractor's offer to the Employer for the execution of the Works.

The use of the word "offer" is unhelpful. The offer consists of a series of documents, not only the financial terms. Maybe "proposal" would be better? Otherwise the different documents forming the proposal (financial, technical etc) should be specified. Also, there is no mention in this definition of a need for the document to be headed or entitled letter of Tender.

1.1.52 "**Local Currency**" means the currency of the Country.

This definition is self-explanatory.

1.1.53 "**Materials**" means things of all kinds (other than Plant), whether on the Site or otherwise allocated to the Contract and intended to form or forming part of the Permanent Works, including the supply-only materials (if any) to be supplied by the Contractor under the Contract.

There is no mention here of Materials which form part of the temporary or underground Works. Thus in underground Works where the temporary ground support may not be part of the Permanent Works (because the Design does not consider it in the final stage), but cannot be compared to scaffolding or formwork because they will remain anyway. Materials should be redefined as follows:

> *"Materials means things of all kinds (other than Plant), whether for use on the Site or otherwise allocated to the Contract and intended to form or forming part of the Permanent Works, but excluding such Materials as are incorporated in the temporary Works, but including supply only Materials (if any) to be supplied by the Contractor under the Contract."*

1.1.54 "**month**" is a calendar month (according to the Gregorian calendar).

This definition as written is incorrect. The key element is the reference to the Gregorian calendar therefore that should not be placed in brackets. In any event this is a rather western way of seeing things, anyway. What about contracts in Islamic countries, or other places with a different calendar?

1.1.55 "**No-objection**" means that the Engineer has no objection to the Contractor's Documents, or other documents submitted by the Contractor under these Conditions, and such Contractor's Documents or other documents may be used for the Works.

This definition is self-explanatory.

1.1.56 "**Notice**" means a written communication identified as a Notice and issued in accordance with Sub-Clause 1.3 [*Notices and Other Communications*].

This is one of the few occasions where it is essential for a document to be headed or entitled "Notice". It is difficult to understand why the definition refers to the written communication as being identified as a Notice. Elsewhere in the definitions Documents are "entitled" which, in fact, means "headed". It is suggested that the same phrase be used in this definition.

1.1.57 "**Notice of Dissatisfaction**" or "**NOD**" means the Notice one Party may give to the other Party if it is dissatisfied, either with an Engineer's determination under Sub-Clause

3.7 [*Agreement or Determination*] or with a DAAB's decision under Sub-Clause 21.4 [*Obtaining DAAB's Decision*].

This definition is clear.

1.1.58 "**Part**" means a part of the Works or part of a Section (as the case may be), which is used by the Employer and deemed to have been taken over under Sub-Clause 10.2 [*Taking Over Parts*].

This definition is nonsensical. The separation of the Works or Section into parts will have taken place at the outset of the execution of the Works. The description is incorrect. A better definition would be:

> *"Part means a part of the Works or of a Section (as the case may be) which for convenience of the Engineer or Employer is separately identified for the purposes of the execution of the Works or for early Taking Over by the Employer."*

1.1.59 "**Particular Conditions**" means the document entitled particular conditions of contract included in the Contract, which consists of Part A – Contract Data and Part B – Special Provisions.

This definition needs to re-focus on the document. Therefore a more accurate definition is:

> *"Particular Conditions means the document headed or entitled Particular Conditions of Contract which consists of part A Contract Data and part B Special Provisions and is included in the Contract."*

1.1.60 "**Party**" means the Employer or the Contractor, as the context requires. "**Parties**" means both the Employer and the Contractor.

This definition is self-explanatory

1.1.61 "**Payment Certificate**" means a payment certificate issued by the Engineer under Clause 14 [*Contract Price and Payment*].

This definition is self-explanatory and note comment to 1.1.42.

1.1.62 "**Performance Certificate**" means the certificate issued by the Engineer (or deemed to be issued) under Sub-Clause 11.9 [*Performance Certificate*].

This definition is self-explanatory.

1.1.63 "**Performance Damages**" means the damages to be paid by the Contractor to the Employer, for the failure to achieve the guaranteed performance of the Plant and/or the Works or any part of the Works (as the case may be), as set out in the Schedule of Performance Guarantees.

The absent element is the requirement for the Employer and/or the Engineer to serve a Notice upon the Contractor that the Contractor is in breach of the requirements of the Guaranteed Performances.

1.1.64 "**Performance Security**" means the security under Sub-Clause 4.2 [*Performance Security*].

This definition is somewhat bald.

There is not any definition of performance security in Sub-Clause 4.2. It is clearly intended to be a document issued by bank or insurance company or other authorised body in a format approved by the Employer to ensure that the Contractor has an independent financial warranty that the Contractor will complete the Contract. It is similar to a bond. As the definition implies there is no reason why an Employer cannot request a Contractor to take on more than one performance security although interestingly the terms of Sub-Clause 4.2 do not appear to allow that request to be made.

A more complete definition would be:

> *"Performance Security means a document issued by bank or insurance company or other authorised body in a format approved by the Employer to ensure that the Contractor has an independent financial warranty that the Contractor will complete the Contract, as required by reason of the authority given to the Employer to require the giving of such security under Sub-Clause 4.2 [Performance Security]."*

1.1.65 "**Permanent Works**" means the works of a permanent nature which are to be executed by the Contractor under the Contract.

This definition is self-explanatory.

1.1.66 "**Plant**" means the apparatus, equipment, machinery and vehicles (including any components) whether on the Site or otherwise allocated to the Contract and intended to form or forming part of the Permanent Works.

This definition is for a Permanent Works encompassed within the definition of Plant or in the alternative where Plant forms a part of the Permanent Works.

1.1.67 "**Programme**" means a detailed time programme prepared and submitted by the Contractor to which the Engineer has given (or is deemed to have given) a Notice of No Objection under Sub-Clause 8.3 [*Programme*].

In this definition there is no need for the descriptors being "detailed time" programme. There is only one programme and that is set out in Sub-Clause 8.3 program. It is only necessary to say:

> *"Programme means the programme prepared and submitted by the Contractor to which the Engineer has given (or is deemed to have given) a Notice of No Objection under Sub-Clause 8.3 [Programme]"*

1.1.68 "**Provisional Sum**" means a sum (if any) which is specified in the Contract by the Employer as a provisional sum, for the execution of any part of the Works or for the supply of Plant, Materials or services under Sub-Clause 13.4 [*Provisional Sums*].

The definition should at least say that it is a sum to remunerate the execution of any part etc.

1.1.69 "**QM System**" means the Contractor's quality management system (as may be updated and/or revised from time to time) in accordance with Sub-Clause 4.9.1 [*Quality Management System*].

Providing the Contractors understand the concept there is no further explanation needed.

Reasonable

It is suggested that within "other definitions" at 1.1.6 the bullet is bitten and a definition of "reasonable" is provided. There are, as the drafters are aware, words which have a vague interpretation. There are references to the words "immediately, prompt and promptly, as soon as practicable, as soon as reasonable" and so on. It appears that the only way to remove doubt and avoid confusion is to add a period of days so as to provide a clear sanction each time these phrases are used. Thus, where there are situations where there is not any sanction then that should be made clear. In the event that a deeming provision or any further implied option to extend time is permitted and encouraged then those particular situations should be made clear.

1.1.70 "**Retention Money**" means the accumulated retention moneys, which the Employer retains under Sub-Clause 14.3 [*Application for Interim Payment*] and pays under Sub-Clause 14.9 [*Release of Retention Money*].

This definition is incorrect.

> *"Retention Money is the percentage of the sums certified on Interim Payment Certificates and held back from payment until the completion of the execution of the Works and the conclusion of the Defects Notification Period."*

There is little point in creating a Contract specific specification defining retention monies in a unique manner.

1.1.71 "**Review**" means examination and consideration by the Engineer of a Contractor's submission in order to assess whether (and to what extent) it complies with the Contract and/or with the Contractor's obligations under or in connection with the Contract.

This definition does not describe the full process.

> *"Review means examination, consideration and, as the case may be, approval, or rejection in whole or in part by the Engineer of a Contractor's submission in order to assess whether and to what extent it complies with the Contract and/or with the Contractor's obligations under or in connection with the Contract."*

The process of review is very important within these Conditions of Contract and it is essential that those reading the definitions easily understand the role and position of the Engineer.

1.1.72 "**Schedules**" means the document(s) entitled schedules prepared by the Employer and completed by the Contractor, as attached to the Letter of Tender and included in the Contract. Such document(s) may include data, lists and schedules of payments and/or rates and prices, and guarantees.

The key point is to recognise that these lists, however they may be composed, do not need to be entitled Schedules providing on Face of the document it is obvious what it is. There is an obsession throughout these definitions to have Documents entitled or headed. The only document which requires to be immediately identified is a Notice.

1.1.73 "**Schedule of Payments**" means the document(s) entitled schedule of payments (if any) in the Schedules showing the amounts and manner of payments to be made to the Contractor.

This definition is important as includes references to payments.

1.1.74 "**Schedule of Performance Guarantees**" means the document(s) entitled schedule of performance guarantees (if any) in the Schedules showing the guarantees required by the Employer for performance of the Works and/or the Plant or any part of the Works (as the case may be), and stating the applicable Performance Damages payable in the event of failure to attain any of the guaranteed performance(s).

However, it is vital that there is a procedure or formula for the Employer to trigger in the event of a Failure in the performance of either the Works, any part of the Works or indeed the Plant.

1.1.75 "**Schedule of Rates and Prices**" means the document(s) entitled schedule of rates and prices (if any) in the Schedules.

As mentioned before there is no need for the document to be headed or entitled Schedule of Rates and Prices providing it is obvious from the contents of the document what it contains. A more appropriate definition may be:

> *"Schedule of Rates and Prices means the Documents containing a schedule of Rates and Prices whether identified in the Contract Data or otherwise."*

1.1.76 "**Section**" means a part of the Works specified in the Contract Data as a Section (if any).

This confusingly refers to a part. A more appropriate definition might be:

> *"Section means a portion of the Works specified or otherwise identified in the Contract Data or elsewhere as a section (if any)."*

1.1.77 "**Site**" means the places where the Permanent Works are to be executed and to which Plant and Materials are to be delivered, and any other places specified in the Contract as forming part of the Site.

Usually in these Conditions opportunities allowed for either plural or singular. This should occur here. Further the reference to the Contract should be widened to include Employer's Requirements and the Contract Data. Thus, a new definition would be:

> *"Site means the place(s) where the Permanent Works are to be executed and to which Plant and Materials are to be delivered and any other places specified in the Contract, as forming part of the Site."*

1.1.78 "**Special Provisions**" means the document (if any), entitled special provisions, which constitutes Part B of the Particular Conditions.

This definition is self-explanatory.

1.1.79 "**Statement**" means a statement submitted by the Contractor as part of an application for a Payment Certificate under Sub-Clause 14.3 [*Application for Interim Payment*], Sub-Clause 14.10 [*Statement at Completion*] or Sub-Clause 14.11 [*Final Statement*].

This definition would be assisted by stating that the Statement means a financial document submitted by the Contractor in support of any application made for payment under this Contract.

1.1.80 "**Subcontractor**" means any person named in the Contract as a subcontractor, or any person appointed by the

Contractor as a subcontractor or designer, for a part of the Works; and the legal successors in title to each of these persons.

The Grammatical punctuation here is inappropriate. The definition should read:

> *"Subcontractor means any person named in the Contract as a Subcontractor or any person appointed by the Contractor as a Subcontractor, supplier or designer for a part of the Works and the legal successors in title to each of these persons."*

This sentence is meant to be read as a single instruction. Breaking the sentence with commas and semicolons detracts from the meaning, which is intended.

1.1.81 "**Taking-Over Certificate**" means a certificate issued (or deemed to be issued) by the Engineer in accordance with Clause 10 [*Employer's Taking Over*].

This definition is self-explanatory.

1.1.82 "**Temporary Works**" means all temporary works of every kind (other than Contractor's Equipment) required on Site for the execution of the Works.

It is difficult to understand how the addition of the phrase in (other than Contractor's equipment) assists the definition. Temporary Works are temporary Works.

1.1.83 "**Tender**" means the Letter of Tender, the Contractor's Proposal, the JV Undertaking (if applicable), and all other documents, which the Contractor submitted with the Letter of Tender, as included in the Contract.

The definition is too wide and too lacking in conditionality. A more practical definition would be as follows:

> *"Tender means the letter of Tender, whether entitled or headed Tender or not, the Contractor's proposal, the JV Undertaking (if any is required) and any other Documents which the Contractor submitted with the letter of Tender as required in the Instructions to Tenderers."*

The key place to find the Schedule of Documents required to be submitted should be in the Employer's Requirements. The identification of any other place is confusing. There is no need to confuse the issue by requiring a Joint Venture Undertaking where there may not be any Joint-Venture.

1.1.84 "**Tests after Completion**" means the tests (if any) which are specified in the Employer's Requirements and which are carried out under Clause 12 [*Tests after Completion*] after the Works or a Section (as the case may be) are taken over under Clause 10 [*Employer's Taking Over*].

This definition ignores the inclusion in these Conditions of Contract of a part of the Works. Therefore the definition should be re-written as follows:

> *"Tests after Completion means the tests (if any) which are specified in the Employer's Requirements and which are carried out under Clause 12 [Tests after Completion] after the Works, a Section or a Part (as the case may be) are taken over by the Employer."*

1.1.85 "**Tests on Completion**" means the tests which are specified in the Contract or agreed by both Parties or instructed as a Variation, and which are carried out under Clause 9 [*Tests on Completion*] before the Works or a Section (as the case may be) are taken over under Clause 10 [*Employer's Taking Over*].

The same concerns with this definition arose with the previous definitions. A consistent redraft would be:

> *"Tests on Completion means the tests which are specified in the Employer's Requirements or agreed by both Parties or instructed as a variation, and which are carried out under Clause 9 [Tests on Completion] before the Works, a Section or a Part as the case may be) are taken over by the Employer."*

1.1.86 "**Time for Completion**" means the time for completing the Works or a Section (as the case may be) under Sub-Clause 8.2 [*Time for Completion*], as stated in the Contract Data as may be extended under Sub-Clause 8.5 [*Extension of Time for Completion*], calculated from the Commencement Date.

Yet again a "part" of the Works is ignored. And a reference to the Employer's Requirements would be more relevant. Therefore a revised definition would be:

> *"Time for Completion means the time for Completing the Works, a Section or a part (as the case may be) under Sub-Clause 8.2 [Time for Completion], as stated in the Employer's Requirements or Contract Data (or as may be extended under Sub-Clause 8.5 [Extension of Time for Completion]), calculated from the Commencement Date."*

1.1.87 "**Unforeseeable**" means not reasonably foreseeable by an experienced contractor by the Base Date.

The addition of the last few words is not helpful. There is a problem with the lapse of time between the Base Date and the Date for Submission of the Tender: in certain cases it is impossible for the Contractor to include information in his tender that appears after base date (a striking example would be some Tsunami on the day before the Date of Submission). This must be clarified, and the Date for Submission is not appropriate.

Therefore the definition should be:

"Unforeseeable means not reasonably foreseeable by an experienced Contractor."

1.1.88 "**Variation**" means any change to the Employer's Requirements or to the Works, which is instructed as a variation under Clause 13 [*Variations and Adjustments*].

This definition is self-explanatory.

1.1.89 "**Works**" mean the Permanent Works and the Temporary Works, or either of them as appropriate.

This definition is self-explanatory.

1.1.90 "**year**" means 365 days.

1.2 Interpretation

In the Contract, except where the context requires otherwise:

(a) words indicating one gender include all genders; and "he", "his" and "himself" shall be read as "he/she", "his/her" and "himself/herself" respectively;
(b) words indicating the singular also include the plural and words indicating the plural also include the singular;

(c) provisions including the word "agree", "agreed" or "agreement" require the agreement to be recorded in writing;

(d) "written" or "in writing" means hand-written, type-written, printed or electronically made, and resulting in a permanent record;

(e) "may" means that the Party or person referred to has the choice of whether to act or not in the matter referred to;

This last is a new interpretation Clause and is useful because it removes any doubt as to whether the word "May" is discretionary or mandatory. Most common sense interpretations indicate the word made to be discretionary but some authoritative institutions consider context decides the meaning in spite of a clear indication to the contrary. This is very much to be welcomed in this interpretation description.

(f) "shall" means that the Party or person referred to has an obligation under the Contract to perform the duty referred to;

The comments made above apply equally to this new interpretation.

(g) "consent" means that the Employer, the Contractor or the Engineer (as the case may be) agrees to, or gives permission for, the requested matter;

There is a question as to whether the reference to "agrees to" is entirely appropriate. A definition in the sense of "Where consent is issued or explicit agreement notified" would be more helpful. There is a request and the response to the request is either in the negative or the positive.

(h) "including", "include" and "includes" shall be interpreted as not being limited to, or qualified by, the stated items that follow;

The phraseology here is inappropriate. A better direction would be as follows:

> *"the words including, include and includes mean 'not limited to' or 'not qualified by' the stated items that follow, as the case may be."*

(i) "words" indicating persons or parties shall be interpreted as referring to natural and legal persons (including corporations and other legal entities); and

This interpretation is not helpful as at least two or three definitions refer to the fact of a person being a natural person. Why is it necessary to define the person as a natural person if the interpretation definition of person here is quite explicit? Therefore, the definition is redundant.

(j) "execute the Works" or "execution of the Works" means the design, construction and completion of the Works and the remedying of any defects.

This definition is self-explanatory.

In any list in these Conditions, where the second-last item of the list is followed by "and" or "or" or "and/or" then all of the list items going before this item shall also be read as if they are followed by "and" or "or" or "and/or" (as the case may be).

The marginal words and other headings shall not be taken into consideration in the interpretation of these Conditions.

Where the condition is included in the first line, which says "except where the context requires otherwise". This condition means that all the elements within this Sub-Clause are subject to this condition. This cannot be right. Take as an example the direction that all provisions that include the word "agree" etc., are required to be recorded in writing, which means that the sentence in question should be put at the beginning of Sub-Clause 1.2.

This condition described above appears to require the users of these Conditions of Contractor consider every paragraph within every Sub-Clause in order to decide whether Sub-Clause 1.2 applies "or whether the context requires otherwise".

As to the second direction stated in this Sub-Clause that the headings etc., shall not be taken into consideration in the interpretation of these Conditions this direction should be re-phrased.

A proposal is:

> *"The headings set out in these Conditions of Contract are merely headings for the provision of guidance and are not taken to have any critical impact upon the contents of the Sub-Clause."*

1.3 Notices and other Communications

Wherever these Conditions provide for the giving of a Notice (including a Notice of Dissatisfaction) or the issuing, providing, sending, submitting or transmitting of another type of communication (including acceptance, acknowledgement, advising, agreement, approval, certificate, Claim, consent, decision, determination, discharge, instruction, No-objection, record(s) of meeting, permission, proposal, record, reply, report, request, Review, Statement, statement, submission or any other similar type of communication), the Notice or other communication shall be in writing and:

This accumulation of nouns and Gerunds is intended to be all encompassing but does not simplify the advice to the Parties. What is required are two simple directions.

> *"Wherever these Conditions provide for the giving of Notice that Notice shall be in writing.*
>
> *Wherever these Conditions provide for the existence of a communication of whatever nature that communication shall be in writing."*

(a) shall be:
   (i) a paper-original signed by the Contractor's Representative, the Engineer, or the authorised representative of the Employer (as the case may be); or
   (ii) an electronic original generated from any of the systems of electronic transmission stated in the Contract Data (if not stated, system(s) acceptable to the Engineer), where the electronic original is transmitted by the electronic address uniquely assigned to each of such authorised representatives,

   or both, as stated in these Conditions; and

This appears to be slightly complex.

In the event that the Parties are not using paper originals then an electronic communication and which is uniquely assigned to each of such authorised Representatives is helpful. Further, there is no requirement here for the Contractor to submit to the Engineer it's proposals for an effective electronic communication system. This is essential in view of the bracketed reference to the system(s) acceptable to the Engineer.

(b) if it is a Notice, it shall be identified as a Notice. If it is another form of communication, it shall be identified as such and include reference to the provision(s) of the Contract under which it is issued where appropriate;

(c) delivered by hand (against receipt), or sent by mail or courier (against receipt), or transmitted using any of the systems of electronic transmission under sub-paragraph (a)(ii) above; and

There is no clear grammatical assistance from the bracketed phrase against receipt with delivery by hand. If it is the intent of the revised Conditions of Contract to have a receipt whenever there is a delivery of a communication by hand then it should be stated explicitly and not encompassed within a bracket.

(d) delivered, sent or transmitted to the address for the recipient's communications as stated in the Contract Data. However, if the recipient gives a Notice of another address, all Notices and other communications shall be delivered accordingly after the sender receives such Notice.

Where these Conditions state that a Notice or NOD or other communication is to be delivered, given, issued, provided, sent, submitted or transmitted, it shall have effect when it is received (or deemed to have been received) at the recipient's current address under sub-paragraph (d) above. An electronically transmitted Notice or other communication is deemed to have been received on the day after transmission, provided no non-delivery notification was received by the sender.

This instruction does not mandate that the electronic transmission system shall have within it software to record receipt and opening of communication. That software is available and regularly used and should be part of the requirement as to delivery set out in sub-paragraph (d).

All Notices, and all other types of communication as referred to above, shall not be unreasonably withheld or delayed.

There are no sanctions applied to situations where a Party has unreasonably withheld or delayed such communication.

When a Notice or NOD or certificate is issued by a Party or the Engineer, the paper and/or electronic original shall be sent to the intended recipient and a copy shall be sent to the Engineer or the other Party, as the case may be. All other communications shall be copied to the Parties and/or the Engineer as stated under these Conditions or elsewhere in the Contract.

This last direction is self-explanatory.

1.4 Law and Language

The Contract shall be governed by the law of the country (or other jurisdiction) stated in the Contract Data (if not stated, the law of the Country), excluding any conflict of law rules.

The ruling language of the Contract shall be that stated in the Contract Data (if not stated, the language of these Conditions). If there are versions of any part of the Contract, which are written in more than one language, the version which is in the ruling language shall prevail.

There could be situations where there is more than one original version of the Contract each written in a different language. Concerns as to interpretation may well arise when operating in a country such as the People's Republic of China where it is highly likely that the Employer will require and original Contract to be written in mandarin side-by-side with are equally binding original Contract written in English. Translations from English into Mandarin and vice a versa are notoriously complex, thus great care must be exercised when operating in such a situation.

Further uncertainty as to which would be the ruling language should be eliminated in the Contract Data in any such situation.

The language for communications shall be that stated in the Contract Data. If no language is stated there, the language for communications shall be the ruling language of the Contract.

The same comments apply.

1.5 Priority of Documents

The documents forming the Contract are to be taken as mutually explanatory of one another. If there is any conflict, ambiguity or discrepancy, the priority of the documents shall be in accordance with the following sequence:

(a) the Contract Agreement;
(b) the Letter of Acceptance;
(c) he Letter of Tender,
(d) the Particular Conditions Part A – Contract Data;
(e) the Particular Conditions Part B – Special Provisions;
(f) these General Conditions;
(g) the Employer's Requirements;
(h) the Schedules;
(i) the Contractor's Proposal;
(j) the JV Undertaking (if the Contractor is a JV); and
(k) any other documents forming part of the Contract.

There are two changes to the priority schedule in the original Yellow Book. They referred to the Particular Conditions part A and part B. However, there is a slight contradiction in separating the two for the purposes as a priority. Definition 1.1.61 refers to the Particular Conditions meaning the document entitled Particular Conditions of Contract. While it is understood that singular means plural and plural means singular and in places in the Conditions such as 1.1.61 it should be made quite clear that the document referred to means Document or Documents.

The second change is in addition to the priority list being the joint-venture Undertaking.

This should be redrafted as the intent of this addition is in the alternative not as a unifying description. That's the reference should be to:

> *"The JV Undertaking, if any (if the Contractor constitutes an unincorporated JV) and (k) any other Documents forming part of the Contract."*

The addition of the alternative option at the end of the sentence makes it quite clear what is required.

If a Party finds an ambiguity or discrepancy in the documents, that Party shall promptly give a Notice to the Engineer, describing the ambiguity or discrepancy. After receiving such Notice, or if the Engineer finds an ambiguity or discrepancy in the documents, the Engineer shall issue the necessary clarification or instruction.

The original direction as to ambiguity was sparse in the extreme. The expanded direction is helpful as it sets out the procedure to be followed. The only comment as to the structure of the direction is that the Engineer might wish to issue a variation, instruction or advice especially where there is a discrepancy. That option is not available to the Engineer. In the last sentence there is a comment "if an ambiguity or discrepancy is found in the Documents, the Engineer shall issue any necessary clarification or instruction". This instruction does not fall under the heading of priority. It is sufficiently important to deserve its own Condition. Further it should make clear that:

> *"where a Party considers that there is an ambiguity or discrepancy of any nature within the Documents then the Party should serve a Notice on the Engineer seeking clarification or instruction which shall be provided by the Engineer within 28 days of receipt. If such a verification or instruction is not provided within the 28 day period then the ambiguity or discrepancy shall be deemed to be found in favour of the Party seeking clarification or instruction."*

1.6 Contract Agreement

The Parties shall sign a Contract Agreement within 35 days after the Contractor receives the Letter of Acceptance, unless they agree otherwise. The Contract Agreement shall be based on the form annexed to the Particular Conditions. The costs of stamp duties and similar charges (if any) imposed by law in connection with entry into the Contract Agreement shall be borne by the Employer.

The revised version of Clause 1.6 refers to the Parties signing a Contract agreement. However, in the last sentence there is a reference to an entry into the Contract Agreement. Therefore, it is suggested that there should be an amendment by removing the words "with entry" and replacing it

> *"in connection with any signing of the Contract Agreement..."*

In the second line of the definition there is a phrase "unless they agree otherwise". It appears that this alternative agreement refers only to the period of days. In fact, it would be quite possible for the Parties to agree not to enter into a Contract Agreement or merely to come to this agreement by implication. An alternative phraseology might be:

> *"unless they agree to vary the entire or any part of the procedure in any way and confirm that variation in writing".*

If the Contractor comprises a JV, the authorised representative of each member of the JV shall sign the Contract Agreement.

Definition 1.1.47 describes the Joint-Venture. Definition 1.1.48 describes in detail a JV Undertaking. It will be usual for the leader of the JV to be authorised to sign the Contract on behalf of the other members of the joint-venture. However, this direction contained in the Contract agreement prevents that situation from occurring.

1.7 Assignment

Neither Party shall assign the whole or any part of the Contract or any benefit or interest in or under the Contract. However, either Party:

(a) may assign the whole or any part of the Contract with the prior agreement of the other Party, at the sole discretion of such other Party; and
(b) may, as security in favour of a bank or financial institution, assign the Party's right to any moneys due, or to become due, under the Contract without the prior agreement of the other Party

This definition has the usual prohibition against assignment. The prohibition is mandatory and absolute. The definition should be amended by a "unless" provision. Thus

"unless the Parties have complied with the permissive Clause set out below".

This addition gives a sensible interpretation to the otherwise total prohibition.

1.8 Care and Supply of Documents

Each of the Contractor's Documents shall be in the custody and care of the Contractor, unless and until submitted to the Engineer. The Contractor shall supply to the Engineer one paper-original, one electronic copy (in the form as specified in the Employer's Requirements or, if not stated, a form acceptable to the Engineer) and additional paper copies (if any) as stated in the Contract Data of each of the Contractor's Documents.

This direction changes the procedure. Previously the Contractor's Documents were retained by the Contractor until the takeover by the Employer. Now these Documents will not remain in the custody and care of the Contractor as soon as they are issued to the Engineer. Since many of the constituent parts of the Contractor's Documents will require review followed by consent or Notice of Objection then the entire procedure has changed and it is suggested appropriately so. Where custody and control lies after the Documents have been issued to the Engineer is not clear.

This procedure has not been thought through. What is intended is that the Documents will be in the custody and care of the Contractor until the execution of the Works has been completed and the Employer has taken over the Works.

There should be a separate instruction directing the Contractor to supply/issue a Document, which will eventually form part of the Contractor's Documents to the Engineer as they come into existence, then Engineer will then review the Documents. That separate provision would then ensure that the Documents remained with the Contractor until completion of the Works. What happens with the reviewed document (who's document is it)? It would be best if everybody kept custody and care of the documents produced by themselves until the end of the Contract

The Contractor shall keep at all times, on the Site, a copy of:

(a) the Contract;
(b) the records under Sub-Clause 6.10 [*Contractor's Records*] and Sub-Clause 20.2.3 [*Contemporary records*];
(c) the publications (if any) named in the Employer's Requirements;
(d) the Contractor's Documents; and
(e) Variations, Notices and other communications given under the Contract.

The Employer's Personnel shall have right of access to all these documents during all normal working hours, or as otherwise agreed with the Contractor.

This set of directions as to the location of Documents is the same as that contained in the original Yellow Book save that the right of access given to the Employer's Personnel has now been restricted to "all normal working hours" as opposed to "at all reasonable times".

This is sensible. There is also flexibility given to the Parties to agree times outside normal working hours.

If a Party (or the Engineer) becomes aware of an error or defect (whether of a technical nature or otherwise) in a document which was prepared by (or on behalf of) the Contractor for use in the execution of the Works, the Party (or the Engineer) shall promptly give a Notice of such error or defect to the other Party (or to the Parties). The Contractor shall then promptly rectify the error or defect at the Contractor's risk and cost.

There are practical changes to this direction from the original Yellow Book. The Engineer can now take the initiative. The error or defect can be technical or otherwise which is very practical.

However, the last sentence is new. That sentence directs that the Contractor shall rectify the error or defect at its own cost. This is reasonable as the errors or defects were found in a document prepared by or on behalf of the Contractor.

The third paragraph addresses the key matter of communication of the nature of an error or defect to the other Party. However, little urgency is given to this instruction; the word used is that the communication must be prompt. There is mention that Notice should be given, however no formality is given with respect to the Notice. No format nor acknowledgement is required! Advice to both Parties is that they should assume formality however laxly the direction is given.

1.9 Errors in the Employer's Requirements

If the Contractor finds an error, fault or defect in the Employer's Requirements as a result of scrutinising them under Sub-Clause 5.1 [*General Design Obligations*], the Contractor shall give a Notice to the Engineer within the period stated in the Contract Data (if not stated, 42 days) calculated from the Commencement Date.

This revised direction is a period of 42 days within which to give a Notice to the Engineer. That was missing from the original direction. It now appears that for reasons which are not clear the revised direction has separated the period into two within which the Notice to be given. The first period is the period is the period stated in the Contract Data calculated from the commencement date and the second period as shall be seen in a moment is any time thereafter.

> If, after expiry of this period, the Contractor finds an error, fault or defect in the Employer's Requirements, the Contractor shall also give a Notice to the Engineer describing the error, fault or defect.

There is no obvious reason for the separation of the period when Notice should be given into two separate periods. Also, there is no opportunity given to the Engineer to find an error, fault or defect in the Employer's Requirements and re-act appropriately by notifying the Parties. There is no timeframe for a Notice in this case. There should be one or at least, there should be "as soon as practicable".

A new element within this direction is as follows:

> The Engineer shall then proceed as specified under Sub-Clause 3.7 [*Agreement or Determination*] to agree or determine:
>
> (a) whether or not there is an error, fault or defect in the Employer's Requirements;
> (b) whether or not (taking account of cost and time) an experienced contractor exercising due care would have discovered the error, fault or other defect:
> - when examining the Site and the Employer's Requirements before submitting the Tender; or
> - if the Contractor's Notice is given after the expiry of the period stated in the first paragraph of this Sub-Clause, when scrutinising the Employer's Requirements under Sub-Clause 5.1 [*General Design Obligations*]; and
>
> (c) what measures (if any) the Contractor is required to take to rectify the error, fault or defect (and, for the purpose of Sub-Clause 3.7.3 [*Time limits*], the date the Engineer receives the Contractor's Notice under this Sub-Clause shall be the date of commencement of the time limit for agreement under Sub-Clause 3.7.3).

What has occurred by way of change is that it appears that the assessment process has now been split into two stages. The first stage is to ascertain whether an experienced Contractor would either during the initial 42-day period or subsequently have discovered the error, fault or other defects and then if so what measures that the Contractor is required to take to rectify the situation.

The 1999 Version of the Yellow Book did not anticipate that the Contractor shall be required to be directed to take appropriate measures. In any event it is sensible to give the Engineer the opportunity to follow through his review of the Notice supplied by the Contractor and if there is agreement that there is a problem then clearly direct the Contractor as to how the problem should be resolved.

If, under sub-paragraph (b) above, an experienced contractor would not have discovered the error, fault or other defect:

(i) Sub-Clause 13.3.1 [*Variation by Instruction*] shall apply to the measures that the Contractor is required to take (if any); and
(ii) if the Contractor suffers delay and/or incurs Cost as a result of the error, fault or defect, the Contractor shall be entitled subject to Sub-Clause 20.2 [*Claims For Payment and/or EOT*] to EOT and/or payment of such Cost Plus Profit.

This last direction to the Engineer is very similar to the direction in the 1999 version.

The key issue will be whether or not the Experienced Contractor would have discovered the error when exercising due care. In any event however experienced the Contractor the Contractor should always serve a Notice whatever the error. Second sight is not available to even an experienced Contractor

If there is a requirement for the Engineer to assess the time and decide whether it is reasonable then the Delay, (and Delay on Site) which can, and will, occur as to what is reasonable and the extent of that reasonability, may be reduced.

It is noted that Sub-Clause 1.9 contains an opportunity for the Contractor to Claim not only an Extension of Time but also payment of any such Costs Plus Reasonable Profit.

1.10 Employer's use of Contractor's Documents

1.10 As between the Parties, the Contractor shall retain the copyright and other intellectual property rights in the Contractor's documents and other design documents made by (or on behalf of) the Contractor

The Contractor shall be deemed (by signing the Contract Agreement) to give to the Employer a non-terminable transferable non-exclusive royalty-free licence to copy, use and communicate the Contractor's Documents and such other design documents, including making and using modifications of them. This licence shall:

This paragraph should be amended as follows:

> *"The Contractor shall be deemed (by signing the Contract) to give the Employer a non-terminable transferable nonexclusive royalty free license to copy, use and communicate the Contractor's Documents and such other Design Documents, including the making and using modifications of them. This license shall:"*

The use of "the" inhibited the effective use of the instruction.

(a) apply throughout the actual or intended operational life (whichever is longer) of the relevant parts of the Works;

The only change here is to remove the reference to working life to operational life.

(b) entitle any person in proper possession of the relevant part of the Works to copy, use and communicate the Contractor's Documents and such other design documents for the purposes of completing, operating, maintaining, altering, adjusting, repairing and demolishing the Works;

The only change here is the addition of the phrase in the second line "and such other Design Documents". This change is compatible with the emphasis within these Conditions of Contract upon the Design requirement of the obligations imposed upon the Contractor.

The area of concern in this Sub-Clause is noted in paragraph 2. In that sentence where the Employer is authorised "to copy, use and communicate" the Documents; those verbs are unusually widened to include making and thereafter using modifications of the Documents.

Two points arise here. The first is that there is no requirement for the Employer to notify the Contractor of each and every modification so that the Contractor can consider the modification and if unwilling to accept the proposal serve a Notice on the Engineer.

Secondly the words used i.e. copy, use and communicate do not of themselves include the making of modifications. This paragraph should be amended to make clear beyond doubt that the Employer is authorised to amend and modify every single part of the Contractor's Documents. The Contractor's Documents are the key elements, which enable the successful construction and frequently operation of the Works. For an Employer to be able to modify them without any formal or even informal communication with the Contractor is not conducive to effective construction of the neither Works nor indeed good relations with the Contractor.

(c) in the case of Contractor's Documents and such other design documents which are in the form of electronic or digital files, computer programs and other software, permit their use on any computer on the Site and/or at the locations of the Employer and the Engineer and/or at other places as envisaged by the Contract; and

Here the changes are in the form of additions. Again, there is a reference to such other Design Documents as was mentioned in the previous paragraph. There is further and more appropriate description of Documents, which are in the form of electronic or digital files. The other additions to the Requirements of the license are that locations of the Employer and Engineer are identified and there is further a requirement for the Contractor and/or the Employer or the Engineer to identify other locations where access to the Documents might be required.

(d) in the event of termination of the Contract:
  (i) under Sub-Clause 15.2 [*Termination for Contractor's Default*], entitle the Employer to copy, use and communicate the Contractor's Documents and the other design documents made by or for the Contractor; or
  (ii) under Sub-Clause 15.5 [*Termination for Employer's Convenience*], Sub-Clause 16.2 [*Termination by Contractor*] or Sub-Clause 18.5 [*Optional Termination*], entitle the Employer to copy, use and communicate the Contractor's Documents for which the Contractor has received payment for the purpose of completing the Works and/or arranging for any other entities to do so.

This element of the application of the license were not dealt with in the earlier Yellow Book version is a matter of practical commonsense in the event of the Termination of the Contract under Sub-Clauses 15.2 and 15.5. It would be very awkward for the Employer to have to litigate with the Contractor in order to be able to obtain permission to use the now terminated Contractor's Documents for the purpose of completing the Works in a situation where the Contract was rightfully terminated.

> The Contractor's Documents and other design documents made by (or on behalf of) the Contractor shall not, without the Contractor's prior consent, be used, copied or communicated to a third party by (or on behalf of) the Employer for purposes other than those permitted under this Sub-Clause.

This is identical to the earlier version.

1.11 Contractor's use of Employer's Documents

> As between the Parties, the Employer shall retain the copyright and other intellectual property rights in the Employer's Requirements and other documents made by (or on behalf of) the Employer. The Contractor may, at the Contractor's cost, copy, use and communicate these documents for the purposes of the Contract.
>
> These documents (in whole or in part) shall not, without the Employer's prior consent, be copied, used or communicated to a third party by the Contractor, except as necessary for the purposes of the Contract.

This Sub-Clause does not permit any modification to be made to these documents. This Sub-Clause is identical to the original Yellow Book version save that there is a reference to the Documents whether whole or part which was omitted from the earlier version. It is difficult to see how the change improves the original version.

1.12 Confidentiality

> The Contractor shall disclose all such confidential and other information as the Engineer may reasonably require in order to verify the Contractor's compliance with the Contract.
>
> The Contractor shall treat all documents forming the Contract as confidential, except to the extent necessary to carry out the Contractor's obligations under the Contract. The Contractor shall not publish, permit to be published, or disclose any

particulars of the Contract in any trade or technical paper or elsewhere without the Employer's prior consent.

The Employer and the Engineer shall treat all information provided by the Contractor and marked "confidential", as confidential. The Employer shall not disclose or permit to be disclosed any such information to third parties, except as may be necessary when exercising the Employer's rights under Sub-Clause 15.2 [*Termination for Contractor's Default*].

A Party's obligation of confidentiality under this Sub-Clause shall not apply where the information:

(a) was already in that Party's possession without an obligation of confidentiality before receipt from the other Party;
(b) becomes generally available to the public through no breach of these Conditions; or
(c) is lawfully obtained by the Party from a third party which is not bound by an obligation of confidentiality.

This Sub-Clause applicable to confidentiality has been extended considerably. There is a sensible direction to the Contractor to treat all Documents which form the Contract as confidential. In addition to that the Contractor is not permitted to gain publicity for itself by disclosing any particulars applicable to the Contract in any place whatsoever whether in writing or orally (although orally is not mentioned) that is implied by the reference to disclosing any particulars of the Contract in any paper or elsewhere.

There is also a useful addition in that the Employer is directed to treat all communications marked (and this is a better phrase than entitled or headed) confidential. The only exception to that embargo is where the Employer is exercising his rights to terminate the Contract under Sub-Clause 15.2 [*Termination for Contractor's Default*].

Then there is a sensible provision which limits confidentiality and removes it in certain circumstances.

The first condition is where the information was already in the Parties' possession where no obligation of confidentiality existed. That is understandable but maybe somewhat difficult to prove. The next option is where the information becomes generally available to the public through no breach of these Conditions. So it refers to some future situation where otherwise the information would remain confidential. Again. the test of how may confidential information become generally available to the public is an interesting test.

Finally, confidentiality may be removed if the information is lawfully obtained from a third Party who has no obligation of confidentiality. Yet again if that circumstance is going to be used effectively then considerable caution must be exercised when relying upon this exception.

This appears to be is a reasonable direction. As the Clause refers to confidential details in the heading it is suggested that it is amended to "confidential and related information or other information".

Further, the description that "the Engineer may reasonably require in order to verify compliance with the Contract" is extremely vague. An alternative draft might be:

> *"the Contractor shall disclose such confidential and other information as the Engineer may from time to time require in order for the Engineer to ensure that the Contractor is complying with all aspects of the Contract".*

Further, such requests must at all times be reasonable and not oppressive.

However, this direction refers to confidential and other information as the Engineer may reasonably require. Though the first question arises as to what may be confidential to the Contractor may be so confidential so that to reveal the information to the Engineer could breach a key confidentiality agreement. This direction does not encourage the Engineer to serve Notices seeking such confidential or other information. There does not appear to the any time limits attached to the requirement of the Contract to disclose the information.

However, there is no opportunity given to the Contractor who seek a Sub-Clause 3.5 [*Determinations*] and Sub-Clause 20.1 [*Contractor's Claim*] determination as to whether the request is reasonable or not and that is the second point in this direction. The Contractor shall only disclose this information where it is reasonable for the Engineer to require that information. Is it for the Contractor to respond to the request, it is not clear whether the request is to be in writing or not, stating why the Engineer is being unreasonable? In the alternative is it for the Engineer to state why it is reasonable for the Engineer to have this information?

Further the Conditions apply to the request as being the only reason for the request can be said so that the Engineer can verify that the Contractor is complying with the Contract. Again, the same questions arise. Is it possible for the Contractor to state that the request does not go to verification? That it goes to another aspect of the Contract? Or that the question relates to matters tangential to the Contract? Again, the question is asked what procedure permits the Contractor to seek a determination that the request does not meet the Requirements of this Sub-Clause other than the usual Sub-Clause 3.5 [*Determinations*] Determination procedure. In the event that the Engineer finds that Sub-Clause 3.5 [*Determinations*] does not apply to those Sub-Clauses being the only opportunity remaining to the Contractor is to seek guidance from the DAAB.

1.13 Compliance with Laws

> The Contractor and the Employer shall, in performing the Contract, comply with all applicable Laws. Unless otherwise stated in the Employer's Requirements:

(a) the Employer shall have obtained (or shall obtain) the planning, zoning or building permit or similar permits, permissions, licences and/or approvals for the Permanent Works, and any other permits, permissions, licenses and/or approvals described in the Employer's Requirements as having been (or being) obtained by the Employer. The Employer shall indemnify and hold the Contractor harmless against and from the consequences of any delay or failure to do so, unless the failure is caused by the Contractor's failure to comply with sub-paragraph (c) below;

This Sub-Clause is identical to that described in the earlier version of the Yellow Book saying that the Employer will be excused any requirement to indemnify the Contractor where the Failure is caused by the Contractor's Failure to comply with sub-paragraph (c)

(b) the Contractor shall give all notices, pay all taxes, duties and fees, and obtain all other permits, permissions, licences and/or approvals, as required by the Laws in relation to the execution of the Works. The Contractor shall indemnify and hold the Employer harmless against and from the consequences of any failure to do so unless the failure is caused by the Employer's failure to comply with Sub-Clause 2.2 [*Assistance*];

This sub-paragraph is also identical to that described in the earlier version of the Yellow Book save that there is an exception given to the requirement to hold harmless where the Employer has failed to play its part in the bargain as required under Sub-Clause 2.2 assistance.

(c) within the time(s) stated in the Employer's Requirement's the Contractor shall provide such assistance and all documentation, as described in the Employer's Requirements or otherwise reasonably required by the Employer, so as to allow the Employer to obtain any permit, permission, licence or approval under sub-paragraph (a) above; and

> (d) the Contractor shall comply with all permits, permissions, licences and/or approvals obtained by the Employer under sub-paragraph (a) above.

Sub-paragraphs (c) and (d) are new. Whether they are particularly applicable under the compliance with Laws sub-paragraph is a matter of debate. It does appear that they are more appropriate to be described under the heading of cooperation.

In any event they require that the Contractor provide assistance where necessary which is quite reasonable. Further, and obviously, sub-paragraph (d) requires the Contractor to comply with all the various permits et cetera which have been properly obtained by the Employer for the execution of the Works.

> If, having complied with sub-paragraph (c) above, the Contractor suffers delay and/or incurs Cost as a result of the Employer's delay or failure to obtain any permit, permission, licence or approval under sub-paragraph (a) above, the Contractor shall be entitled subject to Sub-Clause 20.2 [*Claims For Payment and/or EOT*] to EOT and/or payment of such Cost Plus Profit.

This is new. This ensures that where the Contractor has suffered Delay and/or incurred costs as a result of a breach by the Employer to comply with its own Requirements to award to the Contractor an EOT and/or payment of such Cost Plus Profit.

> If the Employer incurs additional costs as a result of the Contractor's failure to comply with:
>
> (i) sub-paragraph (c) above; or
> (ii) sub-paragraph (b) or (d) above, provided that the Employer shall have complied with Sub-Clause 2.2 [*Assistance*], the Employer shall be entitled subject to Sub-Clause 20.2 [*Claims For Payment and/or EOT*] to payment of these costs by the Contractor.

All this paragraph sets out is where the reverse of the situation has occurred that was anticipated in the previous paragraph and the Contractor has failed to comply with the various Requirements then the Employer may seek reimbursement of its costs which it has incurred as a result of that Failure.

The immediate concern is the Failure of any need for promptness or timeliness on the part of either Party. What sanctions exist other the vague nonsense of being held harmless?

Sub-Clause 1.13(a) states that the Employer shall indemnify and hold the Contractor harmless against and or from the consequences of any Failure to comply with the Conditions of the Sub-Clause. However, there is no reference in this Sub-Clause to Sub-Clauses 1.4 [*Law and Language*]; 1.13 [*Compliance with Laws*]; 3.5 [*Determinations*]; 6.4 [*Labour Laws*] and 20.1 [*Contractor's Claims*]. Nor is there any reference to Sub-Clause 8.4. Nor is this key Clause identified in Sub-Clause 16, which addresses Termination by the Contractor. Where the Employer has failed to comply with the requirement to obtain any relevant permissions then the Contractor should be entitled, at least, to suspend the operation and, in the worst case, to terminate the execution of the Works.

This is a very dangerous Sub-Clause as far as an unwary or inexperienced Contractor is concerned. It could proceed with the Works only to find that operations could be suspended without fault.

Sub-Clause 1.13 (a) further requires the Employer to obtain the planning, zoning and all similar permissions for the Permanent Works.

Then Sub-Clause 1.13 (b) requires the Contractor to obtain all permits, licenses and approvals ...

It appears to the impartial observer that there could well be confusion as to who or which Party should be required to seek approvals for and obtain permission for such approvals as building regulation permission. Normally such permission would be sought by the Party seeking planning permission. However, the phraseology of these Sub-Clauses could indicate that while the Employer shall seek approval for all planning, zoning and similar permissions that detailed building regulation approvals shall be sought by the Contractor. This Section should be clarified.

1.14 Joint and Several Liability

If the Contractor is a Joint Venture:

(a) the members of the JV shall be jointly and severally liable to the Employer for the performance of the Contractor's obligations under the Contract;
(b) the JV leader shall have authority to bind the Contractor and each member of the JV; and
(c) neither the members nor (if known) the scope and parts of the Works to be carried out by each member nor the legal status of the JV shall be altered without the prior consent of the Employer (but such consent shall not

relieve the altered JV from any liability under sub-paragraph (a) above).

Other than the last sentence this Sub-Clause is identical to that encapsulated in the older version of the Yellow Book. Also, there are questions raised as to why the reference to joint and several liability is limited to a joint-venture where previously the scope was wider and included a consortium or other grouping of two or more persons. It appears that the new version of the Yellow Book does not envisage anything other than a joint-venture. However, at definition 1.1.47 there is a definition of joint-venture which now includes the wider scope originally identified in joint and several liabilities that including an association, consortium or other grouping of two or more persons.

It might well be that the law applicable to joint-ventures is different from that applicable to other forms of association. If that is the case then the Employer should ensure that such precautions are taken so was to encompass the association within the definition set out at 1.1.47.

1.15 Limitation of Liability

Neither Party shall be liable to the other Party for loss of use of any Works, loss of profit, loss of any contract or for any indirect or consequential loss or damage which may be suffered by the other Party in connection with the Contract, other than under:

(a) Sub-Clause 8.8 [*Delay Damages*];
(b) sub-paragraph (c) of Sub-Clause 13.3.1 [*Variation by Instruction*];
(c) Sub-Clause 15.7 [*Payment after Termination for Employer's Convenience*];
(d) Sub-Clause 16.4 [*Payment after Termination by Contractor*];
(e) Sub-Clause 17.3 [*Intellectual and Industrial Property Rights*];
(f) the first paragraph of Sub-Clause 17.4 [*Indemnities by Contractor*]; and
(g) Sub-Clause 17.5 [*Indemnities by Employer*].

The total liability of the Contractor to the Employer under or in connection with the Contract, other than:

(i) under Sub-Clause 2.6 [*Employer-Supplied Materials and Employer's Equipment*];
(ii) under Sub-Clause 4.19 [*Temporary Utilities*];
(iii) under Sub-Clause 17.3 [*Intellectual and Industrial Property Rights*]; and
(iv) under the first paragraph of Sub-Clause 17.4 [*Indemnities by Contractor*], shall not exceed the sum stated in the Contract Data or (if a sum is not so stated) the Accepted Contract Amount.

This Sub-Clause shall not limit liability in any case of fraud, gross negligence, deliberate default or reckless misconduct by the defaulting Party.

This Sub-Clause sensibly confirms the extent of liability. There is a slight issue concerning the situation where the acts or inactions are the default of a third party with the knowledge or imputed knowledge of a Party. There should be an additional provision, which allows for liability to extend to those third party actors parties.

1.16 Contract Termination

Subject to any mandatory requirements under the governing law of the Contract, termination of the Contract under any Sub-Clause of these Conditions shall require no action of whatsoever kind by either Party other than as stated in the Sub-Clause.2

This Sub-Clause emphasises that termination shall only take place arising out of and under these Conditions of Contract.

CHAPTER 2

Clause 2 The Employer

2.1 Right of Access to the Site

> The Employer shall give the Contractor right of access to, and possession of, all parts of the Site within the time (or times) stated in the Contract Data. The right and possession may not be exclusive to the Contractor. If, under the Contract, the Employer is required to give (to the Contractor) possession of any foundation, structure, plant or means of access, the Employer shall do so in the time and manner stated in the Employer's Requirements. However, the Employer may withhold any such right or possession until the Performance Security has been received.
>
> 2.1 Right of access to the Site
>
> The Employer shall give the Contractor right of access to, and possession of, or parts of the Site within the time (or times) stated in the Contract Data. The right and possession may not be exclusive to the Contractor. If, under the Contract, the Employer is required to give (to the Contractor) possession of any foundation, structure, Plant or means of access, the Employer shall do so in that time and manner stated in the Employer's Requirements. However, the Employer may withhold any such right or possession until the Performance Security has been received.

Giving access to the Contractor of the Site in a timely and cooperative manner is the key to a successful execution of the Works. The reference to the times stated in the Contract Data from giving access and/or possession is appropriate. Situations can arise where Contractors are not given access to the Site where there has been a misunderstanding or lack of clarity as to which Party is to coordinate the obtaining of access.

It is unfortunate for all concerned to give possession of the Site if access is limited or prohibited. In that situation the Contractor is left in the terrible position of being unable to mobilise his team and commence execution of the Works. If the Contract Data is silent then there is onus on the Employer to provide information to the Contractor as to when and how access and then possession will be given. If there is any question in the Contractor's mind as to when any of these events is likely to take place then the Contractor should issue Notices to the Employer and the Engineer setting out his concerns and making clear that any suggested date for completion may be put at risk. The general Conditions of Contract suggest that position may not be exclusive to the Contractor.

There is no indication in these Conditions of Contract that the Employer should serve a Notice on the Contractor making clear the extent and effect of the reduction in exclusivity. Also, the Employer should serve a Notice in any event should it intend not to give exclusive possession to the Contractor. The Notice should be served at the same time, or earlier, than the giving of the Letter of Acceptance. It is essential that the Contractor be made aware of any such restrictions as they may seriously affect the programme. The Conditions of Contract then refer in this paragraph to a series of handovers relevant to possession of foundations et cetera at the time and manner stated in the Employer's Requirements. The first comment is that these details should be capable of being found in one document; at present there are references to Contract Data and the Employer's Requirements.

Therefore this paragraph should be amended as follows:

> *"The Employer shall give to the Contractor right, or rights, of access to and possession of all parts of the Site within the time, or times, stated in the Employer's Requirements, Contract Data, Contract, but look at the Conditions or any other relevant document. The right to access and to possession may not be exclusive to the Contractor; where the possession is not exclusive the Employer will serve a Notice on the Contractor together with the Letter of Acceptance stating how and in what manner possession will not be exclusive. If the Employer is required to give to the Contractor possession of any foundation, structure, Plant or means of access the Employer shall do so at the times and in the manner stated in the Employer's Requirements or else where as defined above. However, the Employer may withhold any such right or possession until the performance security have been received."*

> If no such time is stated in the Contract Data, the Employer shall give the Contractor right of access to, and possession of, those parts of the Site within such times as may be required to enable the Contractor to proceed in accordance with the Programme or, if there is no Programme at that time, the initial programme submitted under Sub-Clause 8.3 [*Programme*].

Paragraph 2 of this Sub-Clause refers to a situation where no times are identified in any of the relevant Documents. In that case the Employer shall work to the initial programme submitted by the Contractor under Sub-Clause 8.3 [*Programme*].

Therefore, it is very important for the Contractor in that initial programme to make very clear when access and possession are required in order for the Contractor to complete the Works in a timely manner.

If the Contractor suffers delay and/or incurs Cost as a result of a failure by the Employer to give any such right or possession within such time, the Contractor shall be entitled subject to Sub-Clause 20.2 [*Claims For Payment and/or EOT*] to EOT and/or payment of such Cost Plus Profit.

In paragraph 3 is set out a very important but contentious right given to the Contractor where access and/or possession have been restricted or prevented. The Contractor must notify the Employer by Notice as soon as there is any indication of a restriction or Delay in the provision of access to not only the Site itself but also to any areas where aspects of the Works ancillary to the main execution of the Works are carried out. For the Contractor it is essential to provide full details of the extent of the inhibition. The Contractor shall continue to provide updated Notices until the Contractor is satisfied that the Employer has complied with the entirety of his obligations under this Sub-Clause.

However, if and to the extent that the Employer's failure was caused by any error or delay by the Contractor, including an error in, or delay in the submission of, any of the applicable Contractor's Documents, the Contractor shall not be entitled to such EOT and/or Cost Plus Profit.

Then appropriately the Sub-Clause refers to a situation where while there has been a Failure on the part of the Employer to provide access and/or possession that where that particular Failure was induced by any error in or Delay in the submission of any of the Contractor's Documents then the Contractor shall not be entitled to an extension of time plus Cost Plus Profit as outlined in the previous paragraph. However, as must be obvious the error or Delay must be of such a nature, so fundamental a nature, that any reasonable Employer would be unable to fulfil its Requirements under this Sub-Clause. It is not sufficient for there to be a trivial error or Delay in order to relieve the Employer of responsibility.

If, under the Contract, the Employer is required to give to the Contractor possession of any foundation, structure, plant or means of access in accordance with Contractor's Documents, the Contractor shall submit such Contractor's Documents to the Engineer in the time and manner stated in the Employer's Requirements.

This last paragraph within Sub-Clause 2.1, which reminds the Contractor to submit such Contractor's Documents to the Engineer in the time and manner stated in the Employer's Requirements should be located at the beginning of the Sub-Clause. Also this sub-paragraph fails to make clear that the Employer is under no obligation to give possession of any foundation, structure, Plant or means of access if the Contractor has not submitted such Contractor's Documents to the Engineer. Nor is it stated that there shall be a review process. It is a paragraph which hangs in the air without effective linkage to the Entire Conditions of Contract.

This paragraph requires revision as follows:

> *"If the Employer is required to give possession to the Contractor of any foundation, structure, Plant or means of access in accordance with or or any of the Contractor's Documents the Contractor shall submit such Contractor's Documents to the Engineer at the time and manner stated in the Employer's Requirements. If no time and manner are stated in the Employer's Requirements then the Contractor shall submit such Documents for review by the Engineer within 14 days of receipt of the Letter of Acceptance. There after the Engineer shall review the Contractor's Documents relevant to this Sub-Clause and within 14 days issue a Notice of No Objection. If no such Notice is issued within 14 days then the Engineer shall be deemed to have given a Notice of Objection."*

2.2 Assistance

If requested by the Contractor, the Employer shall promptly provide reasonable assistance to the Contractor so as to allow the Contractor to obtain:

(a) copies of the Laws of the Country which are relevant to the Contract but are not readily available; and
(b) any permits, permissions, licences or approvals required by the Laws of the Country (including information required to be submitted by the Contractor in order to obtain such permits, permissions, licences or approvals):
  (i) which the Contractor is required to obtain under Sub-Clause 1.13 [*Compliance with Laws*];
  (ii) for the delivery of Goods, including clearance through customs; and
  (iii) for the export of Contractor's Equipment when it is removed from the Site.

This Sub-Clause appears to focus on to possibly separate category of laws. First the Sub-Clause addresses laws, which are relevant to the Contract, which would not readily be available. There is no indication given as to how or why the lack of availability occurs. If the law is not readily available then arguably it should not be relevant to the execution of the Works. The second category of laws applies to the much more general area of regulations under which comes the Contractor's requirement to apply for various permits, licenses and approvals.

It is noted that here the requirement imposed upon the Employer is mandatory. However, nowhere is there indicated any form of sanction against the Employer for breach of this Sub-Clause. Presumably Notices must be issued promptly by the Contractor, not only requesting the assistance, but subsequently if the assistance is not available then signifying that situation. In spite of the importance of obtaining permissions, licenses and approvals for the effective execution of the Works it is noted that Sub-Clause 16.1 [*Contractor's entitlement to Suspend Work*] and Sub-Clause 16.2 [*Termination by Contractor*] which identify situations where the Contractor may suspend and ultimately terminate the Contract such rights are not available to the Contractor under these Sub-Clauses.

And sub-paragraph (b)(iii) emphasis is placed upon assistance been given to export the Contractor's equipment when it is removed from Site. However, there is an equally important scenario when the Contractor requires to bring equipment into the country in order to execute the Works. There is mention of delivery of goods. Also, this Sub-Clause does not refer to the Contractor's Materials there is merely refers to equipment.

A proposed revision is as follows:

> *"For the removal of the Contractor's Equipment and Materials as the case may be, from the Site when the Engineer authorises such removal."*

A Contractor properly advised may wish to amend the Contract Data, Employer's Requirements et cetera to include an amendment to Sub-Clause 16.1 [*Contractor's Right to Suspend*] to take account of the effect of the Employer's non-compliance with Sub-Clauses 2.1 [*Right of Access to the Site*] and 2.2 [*Assistance*].

2.3 Employer's Personnel and Other Contractors

The Employer shall be responsible for ensuring that the Employer's Personnel and the Employer's other contractors (if any) on or near the Site:

(a) co-operate with the Contractor's efforts under Sub-Clause 4.6 [*Co-operation*]; and

(b) comply with the same obligations which the Contractor is required to comply with under sub-paragraphs (a) to (e) of Sub-Clause 4.8 [*Health and Safety Obligations*] and under Sub-Clause 4.18 [*Protection of the Environment*].

The Contractor may require the Employer to remove (or cause to be removed) any person of the Employer's Personnel or of the Employer's other contractors (if any) who is found, based on reasonable evidence, to have engaged in corrupt, fraudulent, collusive or coercive practice.

The heading is misleading as in fact the Sub-Clause refers to matters relating to cooperation.

This requirement imposed on the Employer is contradictory in that the definition of the Employer's Personnel does not include the Employer's other Contractors. The definition of Employer's Personnel at Sub-Clause 1.1.2.6 does not make any mention of any Subcontractor. Yet Sub-Clause 1.1.2.7 specifically mentions under the heading of Contractor's Personnel each Subcontractor. It might be preferable for Sub-Clause 2.3 to be amended to comply with the definition as stated in Sub-Clause 1.1.2.6.

The amendment proposed is as follows:

> *"The Employer shall be responsible for ensuring that the Employer's Personnel, Employer's nominated Subcontractors and Employer's other Contractors lawfully on-Site ..."*

Sub-Clause 4.6 requires the Contractor to allow appropriate opportunities to both the Employer's Personnel and other Contractor's employed by the Employer to carry out work in the execution on or near the site, which is not included in the Contract without stating how this requirement for cooperation is to be achieved. In sub-paragraph (b) it is implied that Personnel and the other Contractors of the Employer shall take actions similar those of the Contractor under Sub-Clause 4.8 [*Safety Procedures*] and 4.18 [*Protection of the Environment*]. This direction appears unwieldy and impractical.

The Contractor should have management of the compliance with all safety regulations, et cetera. Employer's Personnel should not take actions which are similar to those of the Contractor or which identically replicate those of the Contractor. In matters of safety procedures it is vital that there is one clear chain of command. That command should be that of the Contractor when it comes to safety. Unless, of course, the Engineer decides that the Contractor is in breach of any regulations with regard to safety. There should be one system not, as suggested by this Sub-Clause, two parallel systems.

> *"Where there are two systems the Engineer shall review the systems within 14 days of the issuance of the Letter of Acceptance or as soon as practical thereafter and shall issue a Notice to the Contractor specifying which system shall be implemented."*

As regularly noted there is no sanction in this Sub-Clause available against the Employer for any breaches. Especially in matters of safety there should be an opportunity for the Contractor where there are serious safety breaches by the Employer's Personnel to suspend the Contract Sub-Clause 16.1 [*Suspension and Termination by Contractor*]. That option is not given to the Contractor under this Sub-Clause. The Contractor cannot ban the Employer's Personnel from site.

The last paragraph authorises the Contractor to request the Employer to remove from the site all actors who are found to have engaged in the schedule of prohibited practices. There no simple procedures for the Contractor to follow for example the serving of a Notice drawing the attention of the Employer upon a possible breach and then requiring the Employer to investigate and arrive at a finding. There is no opportunity for input by the Contractor into the investigation process.

2.4 Employer's Financial Arrangements

> The Employer's arrangements for financing the Employer's obligations under the Contract shall be detailed in the Contract Data.
>
> If the Employer intends to make any material change (affecting the Employer's ability to pay the part of the Contract Price remaining to be paid at that time as estimated by the Engineer) to these financial arrangements, or has to do so because of changes in the Employer's financial situation, the Employer shall immediately give a Notice to the Contractor with detailed supporting particulars.

This first paragraph addresses changes. There is now an upfront requirement for the Employer to set out how it will initially finance the Contract in the Contract Data. Thereafter the second paragraph requires the Employer to inform the Contractor immediately should there be any changes in the financial arrangements due to the financial situation or material changes which the Employer himself wishes to make.

This is a considerable improvement on the previous requirement, which merely required the Employer to give such information that was necessary after a period of Notice. There are now to clear obligations placed upon the Employer to be very proactive in providing information to the Contractor in the event that some crisis has occurred in the Employer's financing arrangements.

If the Contractor:

(a) receives an instruction to execute a Variation with a price greater than ten percent (10%) of the Accepted Contract Amount, or the accumulated total of Variations exceeds thirty percent (30%) of the Accepted Contract Amount;
(b) does not receive payment in accordance with Sub-Clause 14.7 [*Payment*]; or
(c) becomes aware of a material change in the Employer's financial arrangements of which the Contractor has not received a Notice under this Sub-Clause, the Contractor may request and the Employer shall, within 28 days after receiving this request, provide reasonable evidence that financial arrangements have been made and are being maintained which will enable the Employer to pay the part of the Contract Price remaining to be paid at that time (as estimated by the Engineer).

These provisions are new and inserted for the assistance of the Contractor. If and when any of the circumstances arise then the Contractor is entitled to request, presumably by Notice, that the Employer provide reasonable evidence that it can meet its outstanding obligations. And the Employer has 28 days in which to do so. The restructuring of this Sub-Clause is positive and helpful as far as the Contractor is concerned.

2.5 Site Data and Items of Reference

The Employer shall have made available to the Contractor for information, before the Base Date, all relevant data in the Employer's possession on the topography of the Site and on sub-surface, hydrological, climatic and environmental conditions at the Site. The Employer shall promptly make available to the Contractor all such data, which comes into the Employer's possession after the Base Date.

The original survey control points, lines and levels of reference (the "items of reference" in these Conditions) shall be specified in the Employer's Requirements or issued to the Contractor by a Notice from the Engineer.

This Sub-Clause is new. It spells out matters, which should form part of the earlier Yellow Book. However, there is no sanction offered to the Contractor in the event that the Employer shall be in breach of its obligations to provide this information. Further the second direction of the Employer to make available to the Contractor such data, which comes into Employer's possession after the Base Date presumably refers to data which amends the original data. There is further reference to the use of the word "promptly" whereas in the previous Sub-Clause direction was made that the information shall be given "immediately".

There is also concern with reference to the requirement that the original survey control points, lines and levels of reference shall be specified in the Employer's Requirements. This Sub-Clause does not say that where these items are not specified in the Employer's Requirements then the Contractor is authorised to issue a Notice to the Engineer seeking such information. The Sub-Clause merely authorises the Engineer to give the Contractor such information. The missing linkage outlined above should be inserted on any amended Sub-Clause 2.5 in order to give practical assistance to the Contractor in such situations.

2.6 Employer-Supplied Materials and Employer's Equipment

If Employer-Supplied Materials and/or Employer's Equipment are listed in the Employer's Requirements for the Contractor's use in the execution of the Works, the Employer shall make such materials and/or equipment available to the Contractor in accordance with the details, times, arrangements, rates and prices stated in the Employer's Requirements.

The Contractor shall be responsible for each item of Employer's Equipment, whilst any of the Contractor's Personnel is operating it, driving it, directing it, using it, or in control of it.

This is a new Sub-Clause. Its contents are practical and helpful.

However, there is no indication as to what procedure should be available to the Contractor in the event that the Employer's Requirements are lacking any of the information supposedly detailed in this Sub-Clause then if there is no information this means that there is no Employer's Material or Equipment. If the information is not detailed enough it is up to the Contractor to ask for specifications during Tender (when the Employer's requirements are already available). Presumably the Contractor must be able to issue a Notice to the Engineer and the Employer requiring the Employer to provide the details, times, arrangements, rates and prices and thereafter make available such Materials and/or equipment so identified. The procedure for effective implementation is lacking

CHAPTER 3

Clause 3 The Engineer

3.1 The Engineer

3.1 The Employer shall appoint the Engineer, who shall carry out the duties assigned to the engineer in the contract.

The Engineer shall be vested with all the authority necessary to act as the Engineer under the Contract.

If the Engineer is a legal entity, a natural person employed by the Engineer shall be appointed and authorised to act on behalf of the Engineer under the Contract.

The Engineer (or, if a legal entity, the natural person appointed to act on its behalf) shall be:

(a) a professional engineer having suitable qualifications, experience and competence to act as the Engineer under the Contract; and
(b) shall be fluent in the ruling language defined in Sub-Clause 1.4 [*Law and Language*].

Where the Engineer is a legal entity, the Engineer shall give a Notice to the Parties of the natural person (or any replacement) appointed and authorised to act on its behalf. The authority shall not take effect until this Notice has been received by both Parties. The Engineer shall similarly give a Notice of any revocation of such authority.

The first question raised is how should the Engineer be vested with all the authority necessary. There is no definition of authority. Nor does the sub-clause direct the Employer to carry out the vesting necessary. There is no opportunity for the Contractor to investigate the extent of the authority. The previous Yellow

Book did not consider the concept of investing saying that the authority would be specified or implied from the Contract.

The earlier version of the Yellow Book referred to the need for the Engineer's staff to include suitably qualified Engineers who are competent to carry out the duties. However, the duties are not specified in this Sub-Clause nor is there any procedure for the Engineer nor the Contractor to ascertain the extent of the duties linked to the qualifications.

However, the new version of the Yellow Book has described the duties separately. They are not specifically linked to the qualifications. The Engineer shall have a suitable qualification and experience and competence in the main Engineering discipline. However, there is no reference yet again to the Employer's Requirements specifying the main Engineering discipline or, if there is more than one, the main Engineering disciplines.

It is of little interest to the Contractor as to whether or not the Engineer has a suitable qualification experience and confidence in the particular Engineering discipline providing that the Engineer knows how to manage the Contract in conjunction with the Contractor. Therefore, the direction as to the qualifications is a direction to the Employer.

The second qualification that the Engineers shall be fluent in the ruling language is sensible.

The second Direction within this Sub-Clause is that the Engineer where it is a legal entity shall give a Notice to the Parties of the natural person appointed and authorised to act on its behalf. That is not explicit English.

The direction should be:

> *"where the Engineer is a legal entity, the Engineer shall give a Notice to the Parties identifying the natural person appointed and authorised to act on its behalf."*

This Sub-Clause also directs that this Notice shall not take effect until it has been received by both Parties. However, there is no requirement for the Parties to acknowledge this vital Notice. The effort to, apparently, improve the Yellow Book that certainly, as far as the first Sub-Clause addressing the Engineer has failed.

3.2 Engineer's Duties and Authority

> Except as otherwise stated in these Conditions, whenever carrying out duties or exercising authority, specified in or implied by the Contract, the Engineer shall act as a skilled professional and shall be deemed to act for the Employer.

It is clear that the structure of this Clause addressing the Engineer is awkward. The active appointment should have been encompassed in the first Sub-Clause. As it would be appropriate there to link the fact of appointment with the qualifications.

Whereas here there is a reference to the duties assigned in the Contract. However, it is suggested that the appropriate place to find the duties would be in the Employer's Requirements.

The intent behind the second direction in this Sub-Clause being that except as otherwise stated in these Conditions the Engineer shall act as a skilled professional and shall be deemed to act for the Employer is unclear. There is no reference here to be occasions where in these Conditions the Engineer would not be carrying out duties or exercising authority specified in or implied by the Contract.

The Engineer shall have no authority to amend the Contract or, except as otherwise stated in these Conditions, to relieve either Party of any duty, obligation or responsibility under or in connection with the Contract.

This direction is clear and needs no comment.

The Engineer may exercise the authority attributable to the Engineer as specified in or necessarily to be implied from the Contract. If the Engineer is required to obtain the consent of the Employer before exercising a specified authority, the requirements shall be as stated in the Particular Conditions. There shall be no requirement for the Engineer to obtain the Employer's consent before the Engineer exercises his/her authority under Sub-Clause 3.7 [*Agreement or Determination*]. The Employer shall not impose further constraints on the Engineer's authority.

This direction gives discretion to the Engineer to exercise the authority attributable to him. The Engineer does not need discretion to exercise the authority but he must exercise the authority attributable to him. The authority is then described in the first sentence as specified in or necessary to be implied from the Contract. However, in the second sentence of this Sub-Clause the reference to authority is linked to being specified in or implied by the Contract. There is no logical reason for the use of different prepositions as the appropriate preposition must be "from" as opposed to "by".

The second sentence within the fourth paragraph directs the Engineer to follow the Requirements stated in the Particular Conditions when seeking the consent of the Employer in certain circumstances. However, there is no fall-back provision here for a situation where the Particular Conditions are silent as to the Requirements.

There is an important direction to the Engineer that there is no requirement for consent to be obtained before The Engineer exercises his authority under Sub-Clause 3.7 [*Agreement or Determination*].

Then there is the statement that the Employer shall not impose further constraints on the Engineer's authority. Further than what is the first question. It is clear that the Sub-Clause anticipate situations where constraints are imposed on the activities of the Engineer.

The Requirements to obtain consent are very different from the situations where consent is required from the Employer. There is no indication whether the situations where the constraints on the authority are to be listed or identified. It is obvious that they should be clearly set out in the Employer's Requirements or the Particular Conditions or both.

> However, whenever the Engineer exercises a specified authority for which the Employer's consent is required, then (for the purposes of the Contract) such consent shall be deemed to have been given.

The fifth paragraph appears to contradict the requirements of the fourth paragraph. Any anticipated situation where, before the exercise of the specified authority, the Employer's consent shall be obtained as is identified above. Yet this fifth paragraph implies that the Engineer does not need to seek consent, whenever the specified authority is exercised, the Employer shall be deemed to have given his consent.

This appears to be contradictory and unhelpful. Further and what is the point in placing the phrase for the purposes of the Contract in brackets. The brackets should be removed in order for the sentence to make some logical sense. In any event as stated earlier it contradicts paragraph 4.

> Any acceptance, agreement, approval, check, certificate, comment, consent, disapproval, examination, inspection, instruction, Notice, No-objection, record(s) of meeting, permission, proposal, record, reply, report, request, Review, test, valuation, or similar act (including the absence of any such act) by the Engineer, the Engineer's Representative or any assistant shall not relieve the Contractor from any duty, obligation or responsibility the Contractor has under or in connection with the Contract.

It is very difficult to make sense of this paragraph six. It appears to say that when a document is issued by the Engineer whatever its implications or explicit directions then in any situation where there is a conflict between the Contractor's duties, obligations or responsibilities and the requirements identified by the "act" under or in connection with the Contract those Contractor's duties, obligations and/or responsibilities shall take precedence.

There must be a procedure implied within this very helpful list of acts which may be carried out by the Engineer for the Contractor to be able to challenge each of those listed acts by means of the issue of a Counter-Notice seeking the exercise of the authority by the Engineer to give a determination and the Sub-Clause 3.7 [*Agreement or Determination*] as to whether the particular act does specifically relieve the Contractor from its otherwise over-riding duties or obligations.

3.3 The Engineer's Representative

> The Engineer may appoint an Engineer's Representative and delegate to him/her in accordance with Sub-Clause 3.4 [*Delegation by the Engineer*] the authority necessary to act on the Engineer's behalf at the Site, except to replace the Engineer's Representative.

This is not clear. The sentence should be split into two to make clear what is required.

"This delegated power shall not include the power to appoint a replacement Representative. That power shall remain with the Engineer".

> The Engineer's Representative (if appointed) shall comply with sub-paragraphs (a) and (b) of Sub-Clause 3.1 [*The Engineer*] and shall be based at the Site for the whole time that the Works are being executed at the Site. If the Engineer's Representative is to be temporarily absent from the Site during the execution of the Works, an equivalently qualified, experienced and competent replacement shall be appointed by the Engineer, and the Contractor shall be given a Notice of such replacement.

In Sub-Clause 3.2 there is emphasis placed upon the need for a Notice appointing a natural person to have been received by all the relevant Parties. The implication is that the Notice should be acknowledged. It would be thought appropriate that the same Requirements apply when the Engineer's Representative is being replaced albeit temporarily.

3.4 Delegation by the Engineer

The Engineer may from time to time assign duties and delegate authority to assistants, and may also revoke such assignment or delegation, by giving a Notice to the Parties, describing the assigned duties and the delegated authority of each assistant. The assignment, delegation or revocation shall not take effect until this Notice has been received by both Parties. However, the Engineer shall not delegate the authority to:

(a) act under Sub-Clause 3.7 [*Agreement or Determination*]; and/or
(b) issue a Notice to Correct under Sub-Clause 15.1 [*Notice to Correct*].

The Engineer may from time to time assign duties and delegate authority to assistants, and may also revoke such assignment or delegation, by giving Notice to the Parties, describing the assigned duties and the delegated authority of each assistant. The assignment, delegation or revocation shall not take effect until the Notice has been received by both Parties. However, the Engineer shall not delegate authority to:

(a) act under Sub-Clause 3.7 [*Agreement or Determination*]; and/or
(b) issue a Notice to Correct under Sub-Clause 15.1 [*Notice to Correct*].

The issue raised in the comment as to the effective acknowledgement of the Notice in every instance relevant to the Engineer or the Engineer's assistant is vital. As it is sufficiently important for the authority not to commence until the Notice has been received receipt is not sufficient and acknowledgement must be explicit.

It is somewhat odd to note that the authority to act under Sub-Clause 3.7 [*Agreement or Determination*] cannot be delegated. Yet the entity exercising that authority may never have been on Site. The Engineer's Representative as described in Sub-Clause 3.3 may have full knowledge of the facts applicable to the situation requiring a determination or agreement, in fact a better knowledge than that of the Engineer. But if the Representative does not have the same qualifications/experience as the Engineer himself, then the provision is better; the Engineer will, anyway, consult with the delegate before issuing a notice.

> Assistants shall be suitably qualified natural persons, who are experienced and competent to carry out these duties and exercise this authority, and who are fluent in the language for communications defined in Sub-Clause 1.4 [*Law and Language*].

It is noted that the more extended Requirements applicable to the role of the Engineer, particular knowledge of one or more of the Engineering disciplines applicable to the Works, does not apply to the assistants.

> Each assistant, to whom duties have been assigned or authority has been delegated, shall only be authorised to issue instructions to the Contractor to the extent defined by the Engineer's Notice of delegation under this Sub-Clause. Any act by an assistant, in accordance with the Engineer's Notice of delegation, shall have the same effect as though the act had been an act of the Engineer.
>
> However, if the Contractor questions any instruction or Notice given by an assistant, the Contractor may by giving a Notice refer the matter to the Engineer. The Engineer shall be deemed to have confirmed the assistant's instruction or Notice if the Engineer does not respond, within 7 days after receiving the Contractor's Notice, reversing or varying the assistant's instruction or Notice (as the case may be).

In paragraph 6 of Sub-Clause 3.2 there is a very long list of acts of the Engineer. Omitted from that list is a response. Either the list should be removed altogether and substituted for one generic noun encompassing communication or the act of response should be added to it.

Any revisions of this Sub-Clause 3.4 should direct the Engineer to respond by Notice in the event that the Engineer wishes to reverse or vary the assistant's instruction or Notice.

3.5 Engineer's Instructions

> The Engineer may issue to the Contractor (at any time) instructions which may be necessary for the execution of the Works, all in accordance with the Contract. The Contractor shall only take

instructions from the Engineer, or from the Engineer's Representative (if appointed) or an assistant to whom the appropriate authority to give instruction has been delegated under Sub-Clause 3.4 [*Delegation by the Engineer*].

Subject to the following provisions of this Sub-Clause, the Contractor shall comply with the instructions given by the Engineer or the Engineer's Representative (if appointed) or delegated assistant, on any matter related to the Contract.

The only comment here is to point out the reference for the assistant in the first paragraph has been carefully spelt out and a clear reference made to Sub-Clause 3.4 [*Delegation by the Engineer*]. However, in the second paragraph there is merely a reference to the delegated assistant. It is always preferable not to change the wording when a mere repetition of the wording in the first paragraph would have been sufficient.

If an instruction states that it constitutes a Variation, Sub-Clause 13.3.1 [*Variation by Instruction*] shall apply.

If not so stated, and the Contractor considers that the instruction:

(a) constitutes a Variation (or involves work that is already part of an existing Variation); or
(b) does not comply with applicable Laws or will reduce the safety of the Works or is technically impossible the Contractor shall immediately, and before commencing any work related to the instruction, give a Notice to the Engineer with reasons. If the Engineer does not respond within 7 days after receiving this Notice, by giving a Notice confirming, reversing or varying the instruction, the Engineer shall be deemed to have revoked the instruction. Otherwise the Contractor shall comply with and be bound by the terms of the Engineer's response.

It should be emphasised that this procedure is the reverse of the procedure described in Clause 3.5. The procedure in 3.5 states that if the Engineer does not respond within seven days of receiving the appropriate Notice then the Engineer shall have been deemed to have confirmed the instruction.

It is not immediately clear why the procedures are so effectively reversed. Both the Engineer and the Contractor need to be alert as to these reverse procedures as both the Engineer and/or the Contractor could be placed at a disadvantage in the event of a misunderstanding of the procedure which, it is suggested, could easily take place.

These carefully thought through but confusing procedures are innovations with regard the Yellow Book. At the next revision procedures should be reconsidered and one model is adopted for both situations.

3.6 Replacement of the Engineer

> If the Employer intends to replace the Engineer, the Employer shall, not less than 42 days before the intended date of replacement, give a Notice to the Contractor of the name, address and relevant experience of the intended replacement Engineer.
>
> If the Contractor does not respond within 14 days after receiving this Notice, by giving a Notice stating an objection to such replacement with reasons, the Contractor shall be deemed to have accepted the replacement.
>
> The Employer shall not replace the Engineer with a person (whether a legal entity or a natural person) against whom the Contractor has raised reasonable objection by a Notice under this Sub-Clause.

The only question raised, in the event of a dispute between the Parties as to what is a reasonable Objection is who shall make that decision. Presumably it cannot be the Engineer.

> If the Engineer is unable to act as a result of death, illness, disability or resignation (or, in the case of an entity, the Engineer becomes unable or unwilling to carry out any of its duties, other than for a cause attributable to the Employer) the Employer shall be entitled to immediately appoint a replacement by giving a Notice to the Contractor with reasons and the name, address and relevant experience of the replacement. This appointment shall be treated as a temporary appointment until this replacement is accepted by the Contractor, or another replacement is appointed, under this Sub-Clause.

This provision is essential for the effective management of the Contract. However, it is suggested that the last sentence of this direction is amended to read:

> *"Or where the Contractor has raised reasonable objection to the replacement proposed another replacement is appointed, under this Sub-Clause."*

This proposed amendment to the drafting removes doubt raised by the repetition of the procedure outlined in the previous paragraph.

This direction is practical except that it does not require the Contractor to give a response within a prescribed period of time.

3.7 Agreement or Determination

> When carrying out his/her duties under this Sub-Clause, the Engineer shall act neutrally between the Parties and shall not be deemed to act for the Employer.
>
> When carrying out his/her duties under this Sub-Clause, the Engineer shall act neutrally between the Parties and shall not be deemed to act for the Employer.

This is an innovation! It is well accepted that the Engineer is the appointee of the Employer. The instruction that the Engineer shall act on a neutral basis will place a burden upon the Engineer who will, of course, naturally attempt to act as an impartial entity but, in fact, will attempt to keep his Employer supportive of his actions.

> Whenever these Conditions provide that the Engineer shall proceed under this Sub-Clause to agree or determine any matter or Claim, the following procedure shall apply:

3.7.1 Consultation to reach agreement

> The Engineer shall consult with both Parties jointly and/or separately, and shall encourage discussion between the Parties in an endeavour to reach agreement. The Engineer shall commence such consultation promptly to allow adequate time to comply with the time limit for agreement under Sub-Clause 3.7.3 [*Time limits*]. Unless otherwise proposed by the Engineer and agreed by both Parties, the Engineer shall provide both Parties with a record of the consultation.

The complexity of these new provisions is astonishing. While the simplicity of the procedure under the original yellow version was to be admired.

This paragraph 3.7.1 requires an Engineer encourage discussion no indication is given as to how this encouragement should be managed.

In the second sentence the Engineer is directed to commence such consultation promptly. However, the first sentence referred to consultation and encouragement of discussion. However, the second sentence does not refer to the discussion process. Therefore the second sentence should be amended as follows:

> *"The Engineer shall commence such consultation and initiate the discussion process promptly..."*

The final sentence is also a grammatical contradiction. Unless otherwise proposed by the Engineer and agreed by both Parties he shall provide both Parties with a record of the consultation. At what point should the Engineer make this proposal? A more affective direction would be:

> *"the Engineer shall provide both Parties with a record of the consultation and discussion process unless one or more of the Parties direct the Engineer not to keep a record."*

It is not right that the Engineer should have the initiative as to whether or not to keep a record of the consultation and discussion process.

> If agreement is achieved within the time limit for agreement under Sub-Clause 3.7.3 [*Time limits*] the Engineer shall give a Notice to both Parties of the agreement, which agreement shall be signed by both Parties. This Notice shall state that it is a "Notice of the Parties' Agreement" and shall include a copy of the agreement.

The first sentence is lacking a noun, it should read:

> *"if agreement is achieved within the time limit stated in Sub-Clause 3.7.3 [Time Limits] the Engineer shall give a Notice containing the heads of agreement to both Parties."*

The next question arises is the direction in the second sentence that the Notice must describe the agreement in detail with supporting Particulars. No conditionality is given to this direction. It is essential that the Parties have the opportunity to agree to settle a dispute and describe the settlement within a very simple form. It may well be that they do not wish to have the agreement described in detail with supporting particulars. It will be sufficient on occasions for both Parties to agree to an EOT without necessarily having to give for reasons.

If:

(a) no agreement is achieved within the time limit for agreement under Sub-Clause 3.7.3 [*Time limits*]; or
(b) both Parties advise the Engineer that no agreement can be achieved within this time limit

whichever is the earlier, the Engineer shall give a Notice to the Parties accordingly and shall immediately proceed under Sub-Clause 3.7.2 [*Engineer's Determination*].

It is noted that there is no procedure for the Parties to advise the Engineer that no agreement can be achieved. Since so much care has been expended in expanding the provisions it would be sensible to require that both Parties to advise the Engineer by Notice that no agreement can be achieved within the time limit.

3.7.2 Engineer's Determination

The Engineer shall make a fair determination of the matter or Claim, in accordance with the Contract, taking due regard of all relevant circumstances.

The Engineer does not require to be directed to make a fair determination as he has already been designated as having a neutral role.

Within the time limit for determination under Sub-Clause 3.7.3 [*Time limits*], the Engineer shall give a Notice to both Parties of his/her determination. This Notice shall state that it is a "Notice of the Engineer's Determination", and shall describe the determination in detail with reasons and detailed supporting particulars.

Here it is perfectly right and proper that an independent determination as in this instance will be required to include reasons and supporting particulars.

In this instance the Parties did not agree as to the solution and both will benefit by learning from the reasons for the independent determination.

3.7.3 Time limits

The Engineer shall give the Notice of agreement, if agreement is achieved, within 42 days or within such other time limit as may be proposed by the Engineer and agreed by both Parties (the "time limit for agreement" in these Conditions), after:

(a) in the case of a matter to be agreed or determined (not a Claim), the date of commencement of the time limit for agreement as stated in the applicable Sub-Clause of these Conditions;
(b) in the case of a Claim under sub-paragraph (c) of Sub-Clause 20.1[*Claims*], the date the Engineer receives a Notice under Sub-Clause 20.1 from the claiming Party; or
(c) in the case of a Claim under sub-paragraph (a) or (b) of Sub-Clause 20.1 [*Claims*], the date the Engineer receives:
  (i) a fully detailed Claim under Sub-Clause 20.2.4 [*Fully Detailed Claim*]; or
  (ii) in the case of a Claim under Sub-Clause 20.2.6 [*Claims of Continuing Effect*], an interim or final fully detailed Claim (as the case may be).

This Sub-Clause is inappropriately phrased. The heading is time limits, however the main content of the Sub-Clause addresses the various forms of presentations to be made by the Contractor. This Sub-Clause should be redrafted.

And the contents set out below should be inserted immediately after the second sentence of 3.7.

> *"The request for a determination or agreement under Sub-Clause 3.7 shall be made either in the form of*
>
> *(a)Contractor's submission for the matter to be agreed or determined as stated in the applicable Sub-Clause of these Conditions; or*
> *(b)in the case of a Claim under sub-paragraph (c) of Sub-Clause 20.1 [Claims], a Notice under Sub-Clause 20.1 from the Claiming Party; or*
> *(c)in the case of a Claim under sub-paragraphs (a) or (b) of Sub-Clause 20.1 [Claims]:*
> *(d)a fully detailed Claim under Sub-Clause 20.2.4 [Fully Detailed Claim]; or in the case of a Claim under Sub-Clause 20.2.6 [Claims of Continuing Effect], the final fully detailed Claim."*

It can be seen that the procedures, which are set out in 3.7 do not apply until the Claim has been submitted for agreement or determination. Therefore, setting out the procedures or formats within the time limit Sub-Clause is inappropriate and unhelpful.

> The Engineer shall give the Notice of his/her determination within 42 days or within such other time limit as may be proposed by the Engineer and agreed by both Parties (the "time limit for determination" in these Conditions), after the date corresponding to his/her obligation to proceed under the last paragraph of Sub-Clause 3.7.1 [*Consultation to reach agreement*].
>
> If the Engineer does not give the Notice of agreement or determination within the relevant time limit:
>
> (i) in the case of a Claim, the Engineer shall be deemed to have given a determination rejecting the Claim; or
> (ii) in the case of a matter to be agreed or determined, the matter shall be deemed to be a Dispute which may be referred by either Party to the DAAB for its decision under Sub-Clause 21.4 [*Obtaining DAAB's Decision*] without the need for a NOD (and Sub-Clause 3.7.5 [*Dissatisfaction with Engineers determination*] and sub-paragraph (a) of Sub-Clause 21.4.1 [*Reference of a Dispute to the DAAB*] shall not apply).

The reference to time limits is indicative of an organisation attempting to impose structure upon the Contract which would be more assisted by reference to time periods. Further the last direction to the Engineer containing the deeming provision as to rejection should include time limits in the plural as opposed to the singular to allow for the various time limits identified above.

3.7.4 Effect of the agreement or determination

> Each agreement or determination shall be binding on both Parties (and shall be complied with by the Engineer) unless and until corrected under this Sub-Clause or, in the case of a determination, it is revised under Clause 21 [*Disputes and Arbitration*].
>
> If an agreement or determination concerns the payment of an amount from one Party to the other Party, the Contractor shall

include such an amount in the next Statement and the Engineer shall include such amount in the Payment Certificate that follows that Statement.

This Sub-Clause, at present, concludes that payment shall be made only by the Employer to the Contractor. There are circumstances within these Conditions of Contract where the Employer might be found to make payment to the Contractor. This opportunity is not provided for in the first two sentences of 3.7.4.

If, within 14 days after giving or receiving the Engineer's Notice of agreement or determination, any error of a typographical or clerical or arithmetical nature is found:

(a) by the Engineer: then he/she shall immediately advise the Parties accordingly; or
(b) by a Party: then that Party shall give a Notice to the Engineer, stating that it is given under this Sub-Clause 3.7.4 and clearly identifying the error. If the Engineer does not agree there was an error, he/she shall immediately advise the Parties accordingly.

In option B is a requirement for a Notice to be given to the Engineer. There is no such requirement for the Engineer to give a Notice to the Parties. That requirement should be replicated in option A.

The Engineer shall within 7 days of finding the error, or receiving a Notice under sub-paragraph (b) above (as the case may be), give a Notice to both Parties of the corrected agreement or determination. Thereafter, the corrected agreement or determination shall be treated as the agreement or determination for the purpose of these Conditions.

This last direction as to a period of seven days does not include the option encompassed in option B for the Engineer to find that that was not an error. The last paragraph should be the subject of an addition as follows:

> *"Where the Engineer has either initiated the Notice or accepted the concept that the Notice containing the identification of the error was correct then the Engineer... et cetera."*

3.7.5 Dissatisfaction with Engineer's determination

> If either Party is dissatisfied with a determination of the Engineer:
>
> (a) the dissatisfied Party may give a NOD to the other Party, with a copy to the Engineer;
> (b) this NOD shall state that it is a "Notice of Dissatisfaction with the Engineer's Determination" and shall set out the reason(s) for dissatisfaction;
> (c) this NOD shall be given within 28 days after receiving the Engineer's Notice of the determination under Sub-Clause 3.7.2 [*Engineer's Determination*] or, if applicable, his/her Notice of the corrected determination under Sub-Clause 3.7.4 [*Effect of the Agreement or Determination*] (or, in the case of a deemed determination rejecting the Claim, within 28 days after the time limit for determination under Sub-Clause 3.7.3 [*Time Limits*] has expired); and
> (d) thereafter, either Party may proceed under Sub-Clause 21.4 [*Obtaining DAAB's Decision*].

These are excellent new provisions, which provide clear steps to be taken

> If no NOD is given by either Party within the period of 28 days stated in sub-paragraph (c) above, the determination of the Engineer shall be deemed to have been accepted by both Parties and shall be final and binding on them.
>
> If the dissatisfied Party is dissatisfied with only part(s) of the Engineer's determination:
>
> (i) this part(s) shall be clearly identified in the NOD;
> (ii) this part(s), and any other parts of the determination that are affected by such part(s) or rely on such part(s) for completeness, shall be deemed to be severable from the remainder of the determination; and
> (iii) the remainder of the determination shall become final and binding on both Parties as if the NOD had not been given.

This provision is another excellent addition.

In the event that a Party fails to comply with an agreement of the Parties under this Sub-Clause 3.7 or a final and binding determination of the Engineer, the other Party may, without prejudice to any other rights it may have, refer the failure itself directly to arbitration under Sub-Clause 21.6 [*Arbitration*] in which case the first and the third paragraphs of Sub-Clause 21.7 [*Failure to Comply with DAAB's Decision*] shall apply to such reference in the same manner as these paragraphs apply to a final and binding decision of the DAAB.

This final direction within the new and/or revised direction for making determinations and/or agreements very sensibly deals with the situations where a Party disagrees with a small yet significant element of the determination.

3.8 Meetings

This is another innovation within the Yellow Book Conditions of Contract.

The Engineer or the Contractor's Representative may require the other to attend a management meeting to discuss arrangements for future work and/or other matters in connection with execution of the Works.

Yet again the addition of a few words would improve this direction.

> *"The Engineer or the Contractor's Representative may require one or the other as appropriate to attend a management meeting to discuss arrangements for future work and/or other matters in connection with the execution of the Works."*

Please note that the redraft has added "the" before "execution".

The Employer's other contractors, the personnel of legally constituted public authorities and/or private utility companies, and/or Subcontractors may attend any such meeting, if requested by the Engineer or the Contractor's Representative.

Again the addition of a few words would improve The structure:
After

"...If requested to do so by the Engineer or Contractor's Representative".

The Engineer shall keep a record of each management meeting and supply copies of the record to those attending and to the Employer. At any such meeting, and in the record, responsibilities for any actions to be taken shall be in accordance with the Contract.

All Parties familiar with meetings know that the second item of business on every agenda is ensuring that the record of the previous meeting is a true record. This Sub-Clause should be amended to add a proviso that:

> *"At the next management meeting any Party attending the previous meeting may propose the corrections and/or amendments to the record thereof. The chair of the meeting shall put the proposal to put to the present members of the meeting and adopt the proposed correction or amendment unless there is a clear majority against such action."*

CHAPTER 4

Clause 4 The Contractor

4.1 Contractor's General Obligations

The Contractor shall execute the Works in accordance with the Contract. When completed, the Works (or Section or Part or major item of Plant, if any) shall be fit for the purpose(s) for which they are intended, as defined and described in the Employer's Requirements (or, where no purpose(s) are so defined and described, fit for their ordinary purpose(s)).

The Contractor shall provide the Plant (and spare parts, if any) and Contractor's Documents specified in the Employer's Requirements, and all Contractor's Personnel, Goods, consumables and other things and services, whether of a temporary or permanent nature, required to fulfil the Contractor's obligations under the Contract.

"Goods" includes Materials. However, the wording is redundant, as Goods imply Equipment, Materials, Plant and Temporary Works.

The Works shall include any work which is necessary to satisfy the Employer's Requirements, Contractor's Proposal and Schedules, or is implied by the Contract, and all works which (although not mentioned in the Contract) are necessary for stability or for the completion, or safe and proper operation, of the Works.

Sometimes these General Conditions mention the Contract on other occasions the Employer's Requirements. In this direction to the Contractor there is every reason to include a direct reference to the Contract as opposed to merely inviting the Contractor to imply something from the Contract.

Thus, a redraft would be:

> *"The Works shall include any work which is necessary to satisfy the Contract or, where applicable, is implied from the Contract,..."*

> The Contractor shall be responsible for the adequacy, stability and safety of all the Contractor's operations and activities, of all methods of construction and of all the Works.
>
> The Contractor shall, whenever required by the Engineer, submit details of the arrangements and methods, which the Contractor proposes to adopt for the execution of the Works. No significant alteration to these arrangements and methods shall be made without this alteration having been submitted to the Engineer.

There is a procedural hiatus in this direction where the Contractor is required to submit details of the arrangements and methods proposed to adopt for the execution of the Works. The submission of these details implies quite clearly that there should be a process of review and subsequent consent, partial consent or rejection or the issuance of a No Objection Notice after the receipt of such detailed information as to the method statements.

Further in the last sentence that is a reference to the requirement that no significant alteration to the arrangements and methods should be made without submitting the proposed alterations to the Engineer. It is noted that the word proposed is missing.

Any proposed alteration of whatever nature should be submitted to the Engineer before it is carried out. The Contractor should state whether in his opinion the alteration is significant or not.

The procedure upon receipt of the Notice of Alteration is also missing. This should be included in order to make sense of this direction.

4.2 Performance Security

> The Contractor shall obtain (at the Contractor's cost) a Performance Security to secure the Contractor's proper performance of the Contract, in the amount and currencies stated in the Contract Data. If no amount is stated in the Contract Data, this Sub-Clause shall not apply.

What is the situation where the Contract Data refers to a Performance Security in specific terms mentions the Currencies but omits the amount? It would be reasonable to imply that the amount is the Contract Price.

For a Contractor to try and argue that by reason of the absence of the amount when all other information was present in the Contract Data was sufficient to remove his obligation to obtain a Performance Security here is unsustainable.

4.2.1 Contractor's Obligations

The first point is obvious being that the first direction in 4.2 is part of the Contractor's obligation. This Sub-Clause should be amended to remove the reference to 4.2 and commence the Sub-Clause at 4.2.1.

> The Contractor shall deliver the Performance Security to the Employer, with a copy to the Engineer, within 28 days after receiving the Letter of Acceptance. The Performance Security shall be issued by an entity and from within a country (or other jurisdiction) to which the Employer gives consent and shall be in the form annexed to the Particular Conditions, or in another form agreed by the Employer (but such consent and/or agreement shall not relieve the Contractor from any obligation under this Sub-Clause).

It is difficult to know exactly what it is intended by the phrase encompassed within the brackets forming the last sentence. If the Performance Security has been provided from within an acceptable country and in the form authorised or agreed by the Employer what other requirement is there for the Contractor to meet?

> The Contractor shall ensure that the Performance Security remains valid and enforceable until the issue of the Performance Certificate and the Contractor has complied with Sub-Clause 11.11 [*Clearance of Site*]. If the terms of the Performance Security specify an expiry date, and the Contractor has not become entitled to receive the Performance Certificate by the date 28 days before the expiry date, the Contractor shall extend the validity of the Performance Security until the issue of the Performance Certificate and the Contractor has complied with Sub-Clause 11.11 [*Clearance of Site*].

Questions raised as to the structure of the Performance Security are contentious. It is very important for the Employer to direct the attention of those drawing up the Employer's Requirements and subsequently the Particular Conditions to set out exactly what the Employer requires as not only suitable countries for issuing the performance security but also acceptable banks to the same end. As this is such a crucial element within the Yellow Book Conditions it is recommended that at the outset the Contractor to give the Employer his proposals as to countries or jurisdictions from which he might seek the Performance Security and entities within both countries or jurisdictions who should or might provide the security.

> Whenever Variations and/or adjustments under Clause 13 [*Variations and Adjustments*] result in an accumulative increase or decrease of the Contract Price by more than twenty percent (20%) of the Accepted Contract Amount:
>
> (a) in the case of such an increase, at the Employer's request the Contractor shall promptly increase the amount of the Performance Security in that currency by a percentage equal to the accumulative increase. If the Contractor incurs Cost as a result of this Employer's request, Sub-Clause 13.3.1 [*Variation by Instruction*] shall apply as if the increase had been instructed by the Engineer; or
> (b) in the case of such a decrease, subject to the Employer's prior consent the Contractor may decrease the amount of the Performance Security in that currency by a percentage equal to the accumulative decrease.

A proposed addition to this direction is as follows:

> *"Whenever variations and or adjustments under Clause 13 [Variations and Adjustments] are relevant to a Section or part of the Works and/or Plant to which a Currency different from the Currency applicable to other elements constituting the Accepted Contract Amount,..."*

Without that introduction this is a very difficult direction to interpret.

4.2.2 Claims under the Performance Security

> The Employer shall not make a Claim under the Performance Security, except for amounts to which the Employer is entitled under the Contract in the event of:
>
> (a) failure by the Contractor to extend the validity of the Performance Security, as described in this Sub-Clause, in which event the Employer may claim the full amount

(or, in the case of previous reduction(s), the full remaining amount) of the Performance Security;

(b) failure by the Contractor to pay the Employer an amount due, as agreed or determined under Sub-Clause 3.7 [*Agreement or Determination*] or agreed or decided under Clause 21 [*Disputes and Arbitration*], within 42 days after the date of the agreement or determination or decision or arbitral award (as the case may be);

(c) failure by the Contractor to remedy a default stated in a Notice given under Sub-Clause 15.1 [*Notice to Correct*] within 42 days or other time (if any) stated in the Notice;

The first two opportunities to Claim are relatively straightforward both for implementation and identification of the sums to be claimed. The request for an amount to be identified by reason of a failure by the Contractor to remedy a default appears more difficult to value without the benefit of the comprehensive procedure. There does not appear to be such a procedure other than in Sub-Clause 7.6.

(d) circumstances which entitle the Employer to terminate the Contract under Sub-Clause 15.2 [*Termination for Contractor's Default*], irrespective of whether a Notice of termination has been given; or

(e) if under Sub-Clause 11.5 [*Remedying of Defective Work off Site*] the Contractor removes any defective or damaged Plant from the Site, failure by the Contractor to repair such Plant, return it to the Site, reinstall it and retest it by the date of expiry of the relevant duration stated in the Contractor's Notice (or other date agreed by the Employer).

The procedure applicable under option (e) is described in Sub-Clause 11.5 [Remedying of Defective Work Offsite]. However, the wording of this option is not clear. The amended text should be as follows:

> *"if under Sub-Clause 11.5 [Remedying of Defective Work Offsite] the Contractor removes any defective or damaged Plant from the Site and the amount of the Performance Security is increased, failure by the Contractor to repair such Plant, return*

it to the Site, reinstall it and re-test it by the expiry date stated in the Contractor's Notice (or other date proposed by the Contractor and agreed by the Employer)."

The Employer shall indemnify and hold the Contractor harmless against and from all damages, losses and expenses (including legal fees and expenses) resulting from a claim under the Performance Security to the extent that the Employer was not entitled to make the Claim.

The bracketed comment states including legal fees and expenses. However, in the main text expenses are clearly identified and in any amendment to these Conditions of Contract reference to expenses within the bracket should be removed.

Any amount which is received by the Employer under the Performance Security shall be taken into account:

(i) in the Final Payment Certificate under Sub-Clause 14.13 [*Issue of FPC*]; or

(ii) if the Contract is terminated, in payment due to the Contractor under Sub-Clause 15.4 [*Payment after Termination for Contractor's Default*], Sub-Clause 15.7 [*Payment after Termination for Employer's Convenience*], Sub-Clause 16.4 [*Payment after Termination by Contractor*], Sub-Clause 18.5 [*Optional Termination*], or Sub-Clause 18.6 [*Release from Performance under the Law*] (as the case may be).

The comma after payment after Termination by Contractor should be removed.

4.2.3 Return of the Performance Security:

The Employer shall return the Performance Security to the Contractor:

(a) within 21 days after the Employer has received his copy of the Performance Certificate and the Contractor has complied with Sub-Clause 11.11 [*Clearance of the Site*]; or

(b) promptly after the date of Termination if the Contract is terminated with Sub-Clause 15.5 [*Termination for Employer's Convenience*], Sub-Clause 16.2 [*Termination by Contractor*], Sub-Clause 18.5 [*Optional Termination*], Sub-Clause 18.6 [*Release from Performance under the Law*].

No comments are required other than the standard comment as to the meaning of promptly.

4.3 Contractor's Representative

The Contractor shall appoint the Contractor's Representative and shall give him/her all authority necessary to act on the Contractor's behalf under the Contract, except to replace the Contractor's Representative.

This wording is a repetition of the earlier wording which is insufficient accurate to represent what is required.
There should be a full stop after Contract. The next sentence should read:

> *"This authority shall not extend to any replacement of the Contractor's Representative."*

The Contractor's Representative shall be qualified, experienced and competent in the main engineering discipline applicable to the Works and fluent in the language for communications defined in Sub-Clause 1.4 [*Law and Language*].

In the definition for Engineer allowance is made for there to be more than one Engineering discipline applicable to the Works. It is considered that there should only be one Contractor's Representative in spite of the possibility that some required knowledge is lacking by the Representative.

Unless the Contractor's Representative is named in the Contract, the Contractor shall, before the Commencement Date, submit to the Engineer for consent the name and particulars of the person the Contractor proposes to appoint as Contractor's Representative. If consent is withheld or subsequently revoked,

> or if the appointed person fails to act as Contractor's Representative, the Contractor shall similarly submit the name and particulars of another suitable replacement for such appointment. If the Engineer does not respond within 28 days after receiving this submission, by giving a Notice to the Contractor objecting to the proposed person or replacement, the Engineer shall be deemed to have given his/her consent.

The reference in the last sentence of this submission is misleading. What this paragraph intends is that it should provide a procedure for the situation when the name of the proposed Contractor's Representative is submitted whether initially or subsequently. The use of the adjective "this" indicates that the procedure only refers to the second or subsequent attempt to obtain approval.

Whether that is a reference to the proposed person or replacement is unclear. In fact, the option for replacement does not apply in this circumstance. On each occasion there is merely a proposed person.

It might be suitable to add in any amendment to these general Conditions:

> *"At this stage the Contractor's Representative shall not have authority on the Site to act on behalf of the Contractor until the expiry of the 28 day period or such lesser period as is convenient to the Engineer to issue a Notice of No Objection."*

> The Contractor shall not, without the Engineer's prior consent, revoke the appointment of the Contractor's Representative or appoint a replacement (unless the Contractor's Representative is unable to act as a result of death, illness, disability or resignation, in which case his/her appointment shall be deemed to have been revoked with immediate effect and the appointment of a replacement shall be treated as a temporary appointment until the Engineer gives his/her consent to this replacement, or another replacement is appointed, under this Sub-Clause).

These Conditions set out in Sub-Clause 4.3 do not provide for a procedure by which the Contractor can inform the Engineer of the imminent revocation of the appointment of the Contractor's Representative. The language proposed is approved because it starts from the point that a replacement should not be the rule. Normally the contractor should not replace his representative at all.

The whole time of the Contractor's Representative shall be given to directing the Contractor's performance of the Contract. The Contractor's Representative shall act for and on behalf of the Contractor at all times during the performance of the Contract, including issuing and receiving all Notices and other communications under Sub-Clause 1.3 [*Notices and Other Communications*] and for receiving instructions under Sub-Clause 3.5 [*Engineer's Instructions*].

The Contractor's Representative shall be based at the Site for the whole time that the Works are being executed at the Site. If the Contractor's Representative is to be temporarily absent from the Site during the execution of the Works, a suitable replacement shall be temporarily appointed, subject to the Engineer's prior consent.

Therefore, there is clearly a procedure for a temporary replacement which must replicate that of a Permanent replacement. That procedure should be equally available where the act is one of revocation.

It is presumed that in the Notice to the Engineer seeking consent for the temporarily absent Representative it will be made clear in the Notice the duration of the absence of the Representative. That is not stated in this Sub-Clause.

The Contractor's Representative may delegate any powers, functions and authority except:

(a) the authority to issue and receive Notices and other communications under Sub-Clause 1.3 [*Notices and Other Communications*]; and
(b) the authority to receive instructions under Sub-Clause 3.5 [*Engineer's Instructions*]),

to any suitably competent and experienced person and may at any time revoke the delegation. Any delegation or revocation shall not take effect until the Engineer has received a Notice from the Contractor's Representative, naming the person, specifying the powers, functions and authority being delegated or revoked, and stating the timing of the delegation or revocation.

All these persons shall be fluent in the language for communications defined in Sub-Clause 1.4 [*Law and Language*].

There is no indication here as to whether or not the Engineer is required to consent or issue a Notice of No Objection. It might be appropriate to add in any amendment that these Notices given under this Sub-Clause shall be for record purposes only.

All these persons shall be fluent in the language for communication defined in Sub-Clause 1.4 [*Law and Language*].

4.4 Subcontractors

The Contractor shall not subcontract:

(a) works with a total accumulated value greater than the percentage of the Accepted Contract Amount stated in the Contract Data (if not stated, the whole of the Works); or
(b) any part of the Works for which subcontracting is not permitted as stated in the Contract Data.

The Contractor shall be responsible for the work of all Subcontractors, for managing and coordinating all the Subcontractors" works, and for the acts or defaults of any Subcontractor, any Subcontractor's agents or employees, as if they were the acts or defaults of the Contractor.

This is a very harsh but understandable obligation placed upon the Contractor. Therefore, the Contractor must attempt to ensure that through insurance and expertly worded Subcontracts the risks are passed down the chain to the Subcontractor's.

The Contractor shall obtain the Engineer's prior consent to all proposed Subcontractors, except:

(i) suppliers of Materials; or
(ii) subcontract for which the Subcontractor is named in the Contract.

Where the Contractor is required to obtain the Engineer's consent to a proposed Subcontractor, the Contractor shall submit the name, address, detailed particulars and relevant experience of such Subcontractor and the work intended to be subcontracted to the Engineer and further information which the Engineer may reasonably require. If the Engineer does not respond within 14 days after receiving this submission (or further information if requested), by giving a Notice objecting to the proposed Subcontractor, the Engineer shall be deemed to have given his/her consent.

It is interesting to note that the Engineer only has 14 days to consider these important relationships with the Employer. Whereas the Engineer has 28 days to consider the merits or otherwise of any Contractor's Representatives. A better way of phrasing this last sentence might be:

> *"If the Engineer does not respond within 14 days after acknowledgement of receipt of this submission or does not within the same period issue a Notice requesting more information then the Engineer shall be deemed to have given his consent to the appointment of the Subcontractor. Where the Engineer has issued a Notice requesting more information then time shall run from the date when the additional information is supplied to and acknowledged by the Engineer."*

The Contractor shall give a Notice to the Engineer not less than 28 days before the intended date of the commencement of each Subcontractor's work, and of the commencement of such work on the Site.

A preferable instruction shall be as follows:

> *"The Contractor shall give Notice to the Engineer, not less than 28 days before the intended date of the commencement of each Subcontractor's work, and a further Notice providing the actual date of commencement of such work on Site."*

4.5 Nominated Subcontractors

In this sub-clause, "nominated subcontractor" means a subcontractor named as such in the employer's requirements or whom the engineer, under sub-clause 13.4 [*provisional sums*], instructs the contractor to employ as a subcontractor

The Subcontractor could be named in another document other than the Employer's Requirements. What then is the status of that Subcontractor and indeed the Contractor vis-a-vis the Subcontractor? It is preferable that a nominated Subcontractor, if nominated before the Contract, can be nominated ONLY in the Employer's Requirements.

4.5.1 Objection to Nomination

The Contractor shall not be under any obligation to employ a nominated Subcontractor whom the Engineer instructs and against whom the Contractor raises reasonable objection by giving a Notice to the Engineer, with detailed supporting particulars, no later than 14 days after receiving the Engineer's instruction. An objection shall be deemed reasonable if it arises from (among other things) any of the following matters, unless the Employer agrees to indemnify the Contractor against and from the consequences of the matter:

- (a) there are reasons to believe that the Subcontractor does not have sufficient competence, resources or financial strength;
- (b) the subcontract does not specify that the Nominated Subcontractor shall indemnify the Contractor against and from any negligence or misuse of Goods by the nominated Subcontractor, the Nominated Subcontractor's agents and employees; or
- (c) the subcontract does not specify that, for the subcontracted work (including design, if any), the Nominated Subcontractor shall:
  - (i) undertake to the Contractor such obligations and liabilities as will enable the Contractor to discharge the Contractor's corresponding obligations and liabilities under the Contract; and
  - (ii) indemnify the Contractor against and from all obligations and liabilities arising under or in connection with the Contract and from the consequences of any failure by the Subcontractor to perform these obligations or to fulfil these liabilities.

It would be preferable to change the requirement to notify the Engineer not more than 14 days after receiving the Instruction. The time for Objection should be time limited and not open-ended. The reference to "among other things" in brackets raises questions. What does this mean? That any other "thing" may give rise to an objection, or that at least one of the following must be present? In the first case the subsequent list is useless, in the second case the language "amongst other things" is useless.

Although it is not spelt out an indemnification from the Employer must identify the areas of concern spelt out by the Contractor and agree to hold the Contractor harmless and without liability from the consequences if the areas concerned are justified. Otherwise these are the normal requirements in the event of a Nominated Subcontract.

4.5.2 Payment to Nominated Subcontractors

The Contractor shall pay to the Nominated Subcontractor the amounts due in accordance with the subcontract. These amounts plus other charges shall be included in the Contract Price in accordance with sub-paragraph (b) of Sub-Clause 13.4 [*Provisional Sums*], except as stated in Sub-Clause 4.5.3 [*Evidence of Payments*].

This is a usual clause defining the payment regime for the Nominated Subcontractor. However, this is a direction that applies to any Subcontractor and thus should be inserted in Sub-Clause 4.4.

4.5.3 Evidence of Payments

Before issuing a Payment Certificate which includes an amount payable to a Nominated Subcontractor, the Engineer may request the Contractor to supply reasonable evidence that the nominated Subcontractor has received all amounts due in accordance with previous Payment Certificates, less applicable deductions for retention or otherwise. Unless the Contractor:

(a) submits this reasonable evidence to the Engineer, or
(b) (i) satisfies the Engineer in writing that the Contractor is reasonably entitled to withhold or refuse to pay these amounts, and

(ii) submits to the Engineer reasonable evidence that the Nominated Subcontractor has been notified of the Contractor's entitlement, then the Employer may (at the Employer's sole discretion) pay, directly to the Nominated Subcontractor, part or all of such amounts previously certified (less applicable deductions) as are due to the Nominated Subcontractor and for which the Contractor has failed to submit the evidence described in sub-paragraphs (a) or (b) above.

However, this is a direction that applies to any Subcontractor and thus should be inserted in Sub-Clause 4.4. This is an awkward subsidiary sentence.

A preferred draft might be:

> *"...then the Employer may (at the Employer's sole discretion) pay to the Nominated Subcontractor part or all of such amounts previously certified, less applicable deductions, as are due to the Nominated Subcontractor and for which the Employer is satisfied that the Contractor has failed to submit sufficient evidence as described in sub-paragraphs (a) or (b) above."*

Thereafter, the Engineer shall give a Notice to the Contractor stating the amount paid directly to the nominated Subcontractor by the Employer and, in the next IPC after this Notice, shall include this amount as a deduction under sub-paragraph (b) of Sub-Clause 14.6.1 [*The IPC*].

There is a hint given in sub-paragraph (b) that the Contractor may withhold payments for some unspecified reasons. These reasons could and should include a situation where the Subcontractor is either dilatory or is totally uncooperative. In which case there should be a requirement for the Contractor to serve a Notice on the Engineer to that effect. This is a flaw in the General Conditions, which has omitted any mechanism to protect the Contractor.

4.6 Co-operation

The Contractor shall, as specified in the Employer's Requirements or as instructed by the Engineer, co-operate with and allow appropriate opportunities for carrying out work by:

(a) the Employer's Personnel;

> (b) any other contractors employed by the Employer; and
> (c) the personnel of any legally constituted public authorities and private utility companies, who may be employed in the carrying out, on or near the Site, of any work not included in the Contract. Such appropriate opportunities may include the use of Contractor's Equipment, Temporary Works, access arrangements, which are the responsibility of the Contractor, and/or other Contractor's facilities or services on the Site.

In the introduction there are references to the instances when co-operation may take place. In the first case the cooperation may be part of the Contract and included in the Contract Price, while in the second case cooperation may not be free of charge.

A matter not made clear in category (c) is that all these Requirements for cooperation apply to Works related to the Permanent Works. It could be interpreted especially when reading the last sentence of the first direction that Works of other nature unrelated to the Permanent Works are being envisaged. That, of course, is not the case. Further, it must be made clear that the activities outlined in (c) envisage works related to or arising out of the Contract and or the execution of the Works. An amendment should be as follows:

> *"The Contractor shall, as specified in the Employer's Requirements or elsewhere, or as instructed by the Engineer, cooperate with and allow all appropriate opportunities for carrying out work which relates to or arises out of the Contract and/or the Permanent Works to."*

The amendment makes explicit that the cooperation is limited to the execution of the Works and not activities, which are unrelated to the Contract and/or the Works.

> The Contractor shall be responsible for the Contractor's construction activities on the Site, and shall use all reasonable endeavours to co-ordinate these activities with those of other contractors to the extent (if any) specified in the Employer's Requirements or as instructed by the Engineer.

This role is more than cooperation it is that of a project manager or joint-venture leader who will be given the responsibility to coordinate all the activities of the other Contractor's. That really has nothing to do with cooperation and it could be argued that the wording, which was appropriate to the first paragraph is inappropriate to the second paragraph.

If the Contractor suffers delay and/or incurs Cost as a result of an instruction under this Sub-Clause, to the extent (if any) that co-operation, allowance of opportunities and coordination was Unforeseeable having regard to that specified in the Employer's Requirements, the Contractor shall be entitled subject to Sub-Clause 20.2 [*Claims For Payment and/or EOT*] to EOT and/or payment of such Cost Plus Profit.

This paragraph is meaningless as drafted. The use of the word "that" in line three is unhelpful. A suggested amendment is as follows:

> *"If the Contractor suffers Delay and/or incurs costs as a result of an instruction under this Sub-Clause, to the extent (if any) that cooperation, allowance of opportunities and coordination was unforeseeable having regard to the requirement for cooperation and allowances for opportunities for carrying out work as specified in the Employer's Requirements or elsewhere, the Contractor shall be entitled subject to Sub-Clause 20.2 [Claims for Payment and/or EOT] to EOT and/or payment of such Costs plus Profit."*

Opportunities can be envisaged which might not have been anticipated in the Employer's Requirements, however, the duties and role of the project manager carrying out Site coordination are fundamental to any properly managed project and it is difficult to anticipate situations when and where increases in the burden related to coordination might be unforeseeable.

The previous version of the Yellow Book emphasised that an Engineer's instruction arising out of this Sub-Clause should constitute a variation in and to the extent that it caused a Contractor to incur unforeseeable cost. That opportunity given to the Engineer to link the instruction as a variation with the concept of unforeseeability has been removed. It could be that the new version deliberately omitted the instruction here to be designated as a variation. Also, the usual Requirements for the Contractor to seek a determination under Sub-Clause 3.7 have been removed here and the Contractor is shown to be entitled to apply directly by Notice under Sub-Clause 20.2 [*Claims for Payment and/or EOT*].

4.7 Setting Out

The Contractor shall set out the Works in relation to the items of reference under Sub-Clause 2.5 [*Site Data and Items of Reference*].

The 1999 Yellow Book ensured that the Contractor had original points, lines and levels of reference which were either specified in the Contract or notified by the Engineer. This has now been removed and it is to be assumed that these various key elements requisite for setting out will be provided under Sub-Clause 2.5 [*Site data and Items of Reference*]. The second paragraph of Sub-Clause 2.5 states that the original survey control points, lines and levels of reference shall be specified in the Employer's Requirements or issued to the Contractor by Notice from the Engineer. There is no reason why for the avoidance of doubt these few key reference points could not be repeated in the Sub-Clause addressing setting out.

4.7.1 Accuracy

> The Contractor shall:
>
> (a) verify the accuracy of all these items of reference before they are used for the Works;
> (b) promptly deliver the results of each verification to the Engineer;
> (c) rectify any error in the positions, levels, dimensions or alignment of the Works; and
> (d) be responsible for the correct positioning of all parts of the Works.

The responsibility placed upon the Employer to ensure the accuracy of the terms of reference has been removed. This is very unhelpful for the Contractor as there is little obligation laid upon the Employer to ensure helpful setting out Data is provided to the Contractor. This is also a severe reversal of the role as it was one of the few matters where the Employer was required to exert some diligence prior to the placing of the Contract.

4.7.2 Errors

> If the Contractor finds an error in any items of reference, the Contractor shall give a Notice to the Engineer describing it:
>
> (a) within the period stated in the Contract Data (if not stated, 28 days) calculated from the Commencement Date, if the items of reference are specified in the Employer's Requirements; or

(b) as soon as practicable after receiving the items of reference, if they are issued by the Engineer under Sub-Clause 2.5 [*Site Data and Items of Reference*].

The earlier version of the Yellow Book did not impose a 28-day or such other period as stated in the Contract Data within which to report an error. That was a good reason for this omission. On various occasions elements of the Site data will or may not be required until after the 28-day period after commencement has passed. However, that totally practical consideration has been removed from the new version of the Yellow Book, and it is suggested to re-include it.

4.7.3 Agreement or determination of rectification measures, delay and/or Cost

After receiving a Notice from the Contractor under Sub-Clause 4.7.2 [*Errors*], the Engineer shall proceed under Sub-Clause 3.7 [*Agreement or Determination*] to agree or determine:

(a) whether or not there is an error in the items of reference;
(b) whether or not (taking account of cost and time) an experienced contractor exercising due care would have discovered such error
- when examining the Site and the Employer's Requirements before submitting the Tender; or
- when scrutinising the Employer's Requirements under Sub-Clause 5.1 [*General Design Obligations*], if the items of reference are specified in the Employer's Requirements and the Contractor's Notice is given after the expiry of the period stated in sub-paragraph (a) of Sub-Clause 4.7.2; and

(c) what measures (if any) the Contractor is required to take to rectify the error (and, for the purpose of Sub-Clause 3.7.3 [*Time limits*], the date the Engineer receives the Contractor's Notice under Sub-Clause 4.7.2 [*Errors*] shall be the date of commencement of the time limit for agreement under Sub-Clause 3.7.3).

There is a complication here as to when the Contractor may give Notice of an error. Sub-Clause 4.7.2 makes clear that where the term of reference is stated in the Employer's Requirements then only the period stated in the Contract Data shall be the window of time within which to report an error. A subsequent period of time applies, but not clearly stated, where the items of reference is issued by the Engineer in the form of instruction. Sub-Clause 4 .7 .3 within sub-paragraph (b) envisages, in the last sentence, an opportunity given to the Contractor, when scrutinising the Employer's Requirements under Sub-Clause 5.1, if the terms of reference are specified in those Requirements and, not or, the Contractor's Notice is given after the expiry of the period dated in sub-paragraph (a) Sub-Clause 4.7.2.

The Contractor is apparently allowed to seek a determination in contradiction to the previous Sub-Clause in spite of the expiry of the time period.

While this is practical and useful it is contradicted by the previous Sub-Clause and Engineer is fully entitled to deem himself governed by Sub-Clause 4.7.2 and exclude anything discovered in the Employer's Requirements after the period set out in the Contract Data.

This new condition been set out in Sub-Clause 4.7.3 requires considerable revision as follows:

> *"After receiving a Notice from the Contractor under Sub-Clause 4.7.2 [Errors], the Engineer shall proceed under Sub-Clause 3.7 [Agreement or Determination] to agree or determine:*
>
> *(a) whether or not there is an error in the items of reference;*
> *(b) thereafter whether or not (taking account of costs and time) an experienced Contractor exercising due care would have discovered such error when examining the Site and the Employer's Requirements before submitting the Tender; or when scrutinising the Employer's Requirements under Sub-Clause 5.1 [General Design Obligations], if the items of reference are specified in the Employer's Requirements; and where the Contractor's Notice is given after the expiry of the period stated in sub-paragraph (a) of Sub-Clause 4.7.2 whether the Delay was justified ; and*
> *(c) What measures (if any) the Contractor is required to take to rectify the error."*

The revision imposes a three-step procedure; first is there an error; then could an experienced Contractor have discovered the error in alternate circumstances and finally whether the Delay in disclosing the error was reasonable.

> If, under sub-paragraph (b) above, an experienced Contractor would not have discovered the error:
>
> (i) Sub-Clause 13.3.1 [*Variation by Instruction*] shall apply to the measures that the Contractor is required to take (if any); and

(ii) if the Contractor suffers delay and/or incurs Cost as a result of the error, the Contractor shall be entitled subject to Sub-Clause 20.2 [*Claims For Payment and/or EOT*] to EOT and/or payment of such Cost Plus Profit.

Sub-paragraph (c) of this Sub-Clause requires that the Contractor to describe the measures which are required to rectify the error. However, the subsequent provision, which addresses the right to issue a variation by instruction applicable to the measures does not apparently give the power to the Engineer to assess whether the measures proposed by the Contractor are appropriate in order to rectify the error.

The last paragraph of this Sub-Clause should be re-ordered. The first conclusion is whether or not an experienced Contractor would have discovered the error, if the experienced Contractor would not have discovered the error would he have suffered Delay and/or incurs costs, if that concept is established then it would be appropriate to examine the measures proposed by the Contractor and ascertain whether they are appropriate in all respects to rectify the error.

4.8 Health and Safety Obligations

The Contractor shall:

(a) comply with all applicable health and safety regulations and Laws;
(b) comply with all applicable health and safety obligations specified in the Contract;
(c) comply with all directives issued by the Contractor's health and safety officer (appointed under Sub-Clause 6.7 [*Health and Safety of Personnel*]);
(d) take care of the health and safety of all persons entitled to be on the Site and other places (if any) where the Works are being executed;
(e) keep the Site, Works (and the other places (if any) where the Works are being executed) clear of unnecessary obstruction so as to avoid danger to these persons;
(f) provide fencing, lighting, safe access, guarding and watching of:

(i) the Works, until the Works are taken over under Clause 10 [*Employer's Taking Over*]; and
(ii) any part of the Works where the Contractor is executing outstanding works or remedying any defects during the DNP; and

(g) provide any Temporary Works (including roadways, footways, guards and fences) which may be necessary, because of the execution of the Works, for the use and protection of the public and of owners and occupiers of adjacent land and property.

One area which is regularly omitted from this schedule of health and safety obligations and yet is crucial to the well-being of the operatives on Site is to direct that the Site is kept tidy and that all Materials and Plant are located in appropriate and well organised locations where possible away from the actual execution of the Works. Such examples include scaffolding, steel reinforcement and shuttering when not in actual use.

Within 21 days of the Commencement Date and before commencing any construction on the Site, the Contractor shall submit to the Engineer for information a health and safety manual which has been specifically prepared for the Works, the Site and other places (if any) where the Contractor intends to execute the Works. This manual shall be in addition to any other similar document required under applicable health and safety regulations and Laws.

It is not clear what is meant by within 21 days of the commencement date. It is assumed that it is intended to mean within 21 days after the commencement date. That implication would support the reference to that the Contractor shall not have commenced any construction on the Site. Again the use of the phrase "any construction" is unhelpful.

It is usual that the first few weeks after the commencement date are very active as far as the Contractor is concerned including erecting hoardings, erecting Site offices, organising the provision of utilities and establishing the veracity or otherwise of all the terms of reference referred to above in Sub-Clause 4.7.2.

Therefore whilst Works of construction per se will not be immediately taking place they could be Works of demolition proceeding which some commentators would define as construction.

The health and safety manual shall set out all the health and safety requirements:

(i) specified in the Employer's Requirements;
(ii) that comply with all the Contractor's health and safety obligations under the Contract; and
(iii) that are necessary to effect and maintain a healthy and safe working environment for all persons entitled to be on the Site and other places (if any) where the Works are being executed.

It is recommended that the second reference should be amended as follows:

> *"That comply with all the Contractor's health and safety obligations under the Contract for the particulars Site and for such areas of land off Site which are used for the execution of the Works."*

This amplification draws attention to the Contractor to the Requirements of the manual must be specifically designed to address the safety Requirements of the particular Site and its Offshoots.

The second suggested revision is to the third requirement as follows:

> *"That are required to effect and maintain a healthy and safe working environment for all persons entitled to be on the Site (including lawful visitors) and other places (if any) where the Works are being executed."*

This manual shall be revised as necessary by the Contractor or the Contractor's health and safety officer, or at the reasonable request of the Engineer. Each revision of the manual shall be submitted promptly to the Engineer.

There is no point in submitting the revisions to the Engineer promptly or at all if the Engineer is not empowered to comment, revise or issue a Notice of No Objection.

In addition to the reporting requirement of sub-paragraph (g) of Sub-Clause 4.20 [*Progress Reports*], the Contractor shall submit to the Engineer details of any accident as soon as practicable after its occurrence and, in the case of an accident causing serious injury or death, shall inform the Engineer immediately.

This haphazard form of reporting is not modern practice. For many years Sites have been required to retain and appoint an accident officer who shall at all times keep an accident record book. There should be an addition to the next revision of these Conditions to include:

> *"The Contractor shall appoint from among his staff a Safety /Accident officer who at all times is responsible for keeping a daily record of accidents of all nature.*
>
> *A copy of this daily report will be provided to the Engineer at the end of each day. The Accident/ Safety officer will in addition report personally to the Engineer or the Engineers Representative immediately in the event that there is an accident causing serious injury or death occurring on Site."*

The Contractor shall, as specified in the Employer's Requirements and as the Engineer may reasonably require, maintain records and make reports (in compliance with the applicable health and safety regulations and Laws) concerning the health and safety of persons and any damage to property.

As has been suggested earlier, this last direction is somewhat vague and the more concise proposals suggested for revisions are more appropriate for major Site operations.

4.9 Quality Management and Compliance Verification Systems

4.9.1 Quality Management System

The Contractor shall prepare and implement a QM System to demonstrate compliance with the requirements of the Contract. The QM System shall be specifically prepared for the Works and submitted to the Engineer within 28 days of the Commencement Date. Thereafter, whenever the QM System is updated or revised, a copy shall promptly be submitted to the Engineer.

There should be a revision here. The first sentence encapsulates the requirement for a quality management system. Then there should follow the elements, which should compose the QM system, which are in fact set out in the second paragraph. Thereafter there should be a direction that it should be specifically prepared for the Works and submitted to the Engineer. The sequence is wrongly juxtaposed.

"The Contractor shall prepare and implement a QM System to demonstrate compliance with the Requirements of the Contract.

The QM System shall be in accordance with the details stated in the Employer's Requirements (if any) and shall include the Contractor's procedures:

(a) To ensure that all Notices and other communications under Sub-Clause 1.3 [Notices and other Communications], Contractor's Documents, reports, as-built records, O&M manuals and contemporary records can be traced, with full certainty, to the Works, Materials, Plant, goods, work (workmanship) or test to which they relate;
(b) to ensure proper coordination and management of interfaces between the stages of execution of the Works, and between Subcontractor's; and
(c) For the submission of Contractor's Documents to the Engineer for review.

The QM System shall be specifically prepared for the Works and submitted to the Engineer within 28 days of the commencement date. Thereafter, whenever the QM system is updated or revised, a copy shall be promptly submitted to the Engineer."

That procedural structure follows the logical path required for effective management of this Sub-Clause.

The QM System shall be in accordance with the details stated in the Employer's Requirements (if any) and shall include the Contractor's procedures:

(a) to ensure that all Notices and other communications under Sub-Clause 1.3 [*Notices and Other Communications*], Contractor's Documents, as-built records, O&M Manuals, and contemporary records can be traced, with full certainty, to the Works, Goods, work, workmanship or test to which they relate;
(b) to ensure proper coordination and management of interfaces between the stages of execution of the Works, and between Subcontractors; and
(c) for the submission of Contractor's Documents to the Engineer for Review.

The Engineer may Review the QM System and may give a Notice to the Contractor stating the extent to which it does not comply with the Contract. Within 14 days after receiving this Notice, the Contractor shall revise the QM System to rectify such non-compliance. If the Engineer does not give such a Notice within 21 days of the date of submission of the QM System, the Engineer shall be deemed to have given a Notice of No-objection.

There is a gap in the procedure. The Engineer has received the QM System. The Engineer identifies areas of non-compliance and gives a Notice to that effect to the Contractor. Within 14 days after receipt of that Notice the Contractor responds with revisions to the QM System. Thereafter the Engineer must have a period of time, say 14 days, to consider the revisions and issue a further Notice, whether that is a Notice seeking further revisions or a Notice of No Objection. The Engineer must be given a fixed period of time to consider the revisions and this should be provided for in subsequent revisions of this Yellow Book.

> The Engineer may, at any time, give a Notice to the Contractor stating the extent to which the Contractor is failing to correctly implement the QM System to the Contractor's activities under the Contract. After receiving this Notice, the Contractor shall immediately remedy such failure.

The words "To his activities under the Contract" are superfluous. All the Engineer is required to do is to point out failings in the implementation of the QM System.

> The Contractor shall carry out internal audits of the QM System regularly, and at least once every six months. The Contractor shall submit to the Engineer a report listing the results of each internal audit within seven days of completion. Each report shall include, where appropriate, the proposed measures to improve and/or rectify the QM System and/or its implementation.
>
> If the Contractor is required by the Contractor's quality assurance certification to be subject to external audit, the Contractor shall immediately give a Notice to the Engineer describing any failing(s) identified in any external audit. If the Contractor is a JV, this obligation shall apply to each member of the JV.

This is the first mention of Quality Assurance Certification. As Quality Assurance Certification is intended to feature in these Conditions of Contract then a separate Sub-Clause should be provided to address that issue, Quality management may not, of itself, require Quality Assurance Certification

4.9.2 Compliance Verification System

The Contractor shall prepare and implement a Compliance Verification System to demonstrate that the design, Materials, Employer-Supplied Materials (if any), Plant, work and workmanship comply in all respects with the Contract.

The Compliance Verification System shall be in accordance with the details stated in the Employer's Requirements (if any) and shall include a method for reporting the results of all inspections and tests carried out by the Contractor. In the event that any inspection or test identifies a non-compliance with the Contract, Sub-Clause 7.5 [*Defects and Rejection*] shall apply.

The Contractor shall prepare and submit to the Engineer a complete set of compliance verification documentation for the Works or Section (as the case may be), fully compiled and collated in the manner described in the Employer's Requirements or, if not so described, in a manner acceptable to the Engineer.

A few concerns arise. The imposition of a compliance verification system is a recent innovation. These Conditions of Contract do not give a hint as to which element of the procedure shall be subject to this compliance verification other than to inspections and tests. FIDIC should consider at least one or more compliance verification systems and subsequently add a guidance note for users of their Contracts indicating that the identified systems have been utilised to address certain areas of Contract management and have been far more effective.

This Sub-Clause further does not put in place a clear procedure to assist both the Contractor and the Engineer in preparing and subsequently reviewing the documentation. The last paragraph directs the Contractor to prepare a complete set of verification documentation for the Works or Section. It may well be as the paragraph anticipates that the Employer's Requirements are silent. If so at what stage does the Contractor meet with the Engineer and determine the proper compilation and collation of the documentation? **The sample table of contents of Employer's Requirements should show this passage.**

There should be a preliminary stage whereby the Contractor will either be able to identify immediately from the Employer's Requirements which system must be put in place or if the Requirements are silent then a strict imposition of a time limit for the provision of the information from the Engineer. Many Contractor's while used to quality measurement and management may not have met the sophistication of compliance verification.

Unless the Employer's Requirements state clearly which compliance verification system is required and for what purpose then both Parties should agree to strike out this Sub-Clause. The uncertainty, which seems to be obvious when considering compliance verification is emphasised when reviewing the next Sub-Clause.

4.9.3 General provision

> Compliance with the QM System and/or Compliance Verification System shall not relieve the Contractor from any duty, obligation or responsibility under or in connection with the Contract.

It is obvious from the way that this Sub-Clause is drafted that FIDIC consider that compliance verification may be used as an alternative to the QM System. As that is the case then it is recommended that this Sub-Clause 4.9.2 is not enforced until there is greater understanding and acceptance of the concepts.

4.10 Use of Site Data

> The Contractor shall be responsible for interpreting all data referred to under Sub-Clause 2.5 [*Site Data and Items of Reference*].
>
> To the extent which was practicable (taking account of cost and time), the Contractor shall be deemed to have obtained all necessary information as to risks, contingencies and other circumstances which may influence or affect the Tender or Works. To the same extent, the Contractor shall be deemed to have inspected and examined the Site, access to the Site, its surroundings, the above data and other available information, and to have been satisfied before submitting the Tender as to all matters relevant to the execution of the Works, including:
>
> (a) the form and nature of the Site, including sub-surface conditions;
> (b) the hydrological and climatic conditions, and the effects of climatic conditions at the Site;
> (c) the extent and nature of the work and Goods necessary for the execution of the Works;
> (d) the Laws, procedures and labour practices of the Country; and

(e) the Contractor's requirements for access, accommodation, facilities, personnel, power, transport, water and any other utilities or services.

Paragraph 2 of this Sub-Clause states to the extent practicable, taking account of costs and time, the Contractor shall be deemed to have obtained all necessary information as to various circumstances which may influence or affect the Tender for the Works.

This is a very key and controversial Sub-Clause. It is the Sub-Clause that places every risk upon the Contractor. The reference to subsurface conditions **should clearly be part of the Employer's responsibility.**

It is very difficult to relate the deeming provision set out in this Sub-Clause being that the Contractor shall be deemed to have inspected and examined a whole range of matters before submitting the Tender to the realities of a situation where much information will only come to light once the Contractor has taken possession of the Site. The Employer, in Sub-Clause 2.5, is required to make available all Site data and relevant information which is in his possession. There is no sanction placed against the Employer for failing to make reasonable efforts to ensure that all such data and other available information was submitted and was available to the Contractor at the Base Date.

There are two aspects of great concern, which are particularly drawn to the Contractor's attention.

The first refers to the Laws, procedures and labour practices of the country. That direction is impliedly linked with Sub-Clause 1.13 [*Compliance with Laws*] and Sub-Clause 1.4 [*Law and Language*]. As will be recalled, the definition of Laws includes regulations, ordinances et cetera. The simplest aspect or otherwise of this direction to the Contractor depends upon whether the country has clearly established rules for regulation of labour or whether the organisation of labour is ad hoc or piecemeal, rather unregulated.

There may be different employment regulations depending upon whether the Employer is local in the country or foreign. In some jurisdictions health and safety regulations are ignored or waived by means of a brown envelope. The Contractor needs as far as possible to familiarise himself with the actual working practices on Site in a country in which the Contractor is nor used to operating. This concern not only extends to labour but also to the availability and cost of utilities.

The final major concern for the Contractor, which is not spelled out in this Sub-Clause, is the need to identify the form and nature of the Site but also of other areas of land which will form part of the Site while the Works are being executed but will be Vacated upon completion. The location, availability, cost and access to these other areas of land are vital pieces of information without which the Contractor may be unable to proceed on Site. Yet they are not identified in this Sub-Clause. One of the drawbacks of this Yellow Book is the limitations described within the Sub-Clause addressing cooperation.

Cooperation does not extend to the Employer using its best endeavours to ensure that matters addressing additional lands and access to those lands for off Site Works and the security and provision of utilities generally are placed at the forefront of the Employer's obligations.

4.11 Sufficiency of the Accepted Contract Amount

The Contractor shall be deemed to:

(a) have satisfied himself/herself as to the correctness and sufficiency of the Accepted Contract Amount; and
(b) have based the Accepted Contract Amount on the data, interpretations, necessary information, inspections, examinations and satisfaction as to all relevant matters described in Sub-Clause 4.10 [*Use of Site Data*] and any further data relevant to the Contractor's design.

Unless otherwise stated in the Contract, the Accepted Contract Amount shall be deemed to cover all the Contractor's obligations under the Contract and all things necessary for the proper execution of the Works in accordance with the Contract.

It is useful to note there is no reference here to the Employer's Requirements.

In Sub-Clause 4.1 there is a very clear reference, in the first paragraph, that the Works shall be as defined and described in the Employer's Requirements. However, it might be thought appropriate to repeat the reference to comply with the purposes intended by the Employer and defined and described in the Requirements. It is necessary to recall that these Conditions of Contract do allow for situations where the Accepted Contract Amount shall be varied.

In any revision of these Conditions of Contract it might be appropriate to begin the second paragraph as follows:

"Unless otherwise stated in these Conditions of Contract and in the Contract ..."

4.12 Unforeseeable Physical Conditions

In this Sub-Clause, "physical conditions" means natural physical conditions and physical obstructions (natural or man-made) and pollutants, which the Contractor encounters at the Site during execution of the Works, including sub-surface and hydrological conditions but excluding climatic conditions at the Site and the effects of those climatic conditions.

This is one of the most important Sub-Clauses within any Conditions of Contract.

The importance placed on transparent disclosure by the Employer of all relevant information and Sub-Clause 2.5 provision of data is fundamental not only to the success or failure of the Contract but to the relationship between the Parties. The Contractor is limited in the scope of his investigation by what is reasonable in terms of time and cost.

Unforeseeable man-made obstacles will include bombs fired or dropped onto a location decades before, unexploded, and for which no local records were recorded.

If the Contractor encounters physical conditions which the Contractor considers to have been Unforeseeable and that will have an adverse effect on the progress and/or increase the Cost of the execution of the Works, the following procedure shall apply:

As always, the key will be the interpretation of Unforeseeable. The second concern will be when the unforeseeable aspects have been approved but whether or not there are adverse issues may not be clear at the outset but may be only evident after a period of time.

4.12.1 Contractor's Notice

After discovery of such physical conditions, the Contractor shall give a Notice to the Engineer, which shall:

(a) be given as soon as practicable and in good time to give the Engineer opportunity to inspect and investigate the physical conditions promptly and before they are disturbed;
(b) describe the physical conditions, so that they can be inspected and/or investigated promptly by the Engineer;

(c) set out the reasons why the Contractor considers the physical conditions to be Unforeseeable; and
(d) describe the manner in which the physical conditions will have an adverse effect on the progress and/or increase the Cost of the execution of the Works.

The requirement to describe the physical Conditions is appropriate but cannot always be done in the fullest extent by reason of the surrounding hostile Conditions.

The final direction is to describe the manner in which the physical Conditions will have an adverse effect on the progress and also the request for a budget as to the increase in cost will be an estimate only at this early stage. Since the direction to the Contractor is that the attention of the Engineer shall be alerted as soon as possible it is likely that progress will be slowed or halted until the Engineer has had an opportunity to carry out the initial inspection, where feasible, and investigation.

4.12.2 Engineer's inspection and investigation

The Engineer shall inspect and investigate the physical conditions within 7 days, or longer period agreed with the Contractor, after receiving the Contractor's Notice.

The Contractor shall continue execution of the Works, using such proper and reasonable measures as are appropriate for the physical conditions and to enable the Engineer to inspect and investigate them.

It is recommended that this direction to the Engineer is amended at the next revision to read as follows:

> *"The Engineer shall (if safe to do so) inspect and investigate the physical Conditions as soon as reasonably practical and in any event within seven days, or such longer period as agreed with the Contractor, after receiving the Contractor's Notice."*

The Contractor shall continue execution of the Works, using such proper and reasonable measures as are appropriate for the physical condition and to enable the Engineer to inspect and investigate the physical Conditions. It is recommended that this direction to the Contractor is revised as follows:

> *"The Contractor shall continue execution of the Works, using such safe, precautionary and reasonable measures as are appropriate for the anticipated physical Conditions so as to enable, where feasible, the Engineer to inspect and investigate the physical Conditions."*

This amended direction to the Contractor allows for the Contractor to take reasonable measures to secure the safe working of the Contractor's Personnel as well as ensuring that the physical Conditions remain accessible, where practical, for the Engineer's inspection and investigation. In such circumstances the Engineer should always alert the Contractor as to his availability and indicate to the Contractor such preliminary measures the Engineer envisages might be necessary.

4.12.3 Engineer's instructions

> The Contractor shall comply with any instructions which the Engineer may give for dealing with the physical conditions and, if such an instruction constitutes a Variation, Sub-Clause 13.3.1 [*Variation by Instruction*] shall apply.

4.12.2 Engineer's Inspection and Investigation

> The Engineer shall (if safe to do so) inspect and investigate visible Conditions within seven days, or longer period agreed with the Contractor, after receiving the Contractor's Notice.

It may well be that in such circumstances the Engineer may make a preliminary determination with which the Contractor does not agree. This situation may well be extremely tense and relationships may be uneasy. It is very important even in a critical situation that the Parties involved, the Contractor and the Engineer, without fail follow the Contractual directions as to the issuance of clear Notices at all times. The additional costs incurred in a situation where there is an unforeseeable physical condition would not be awarded to the Contractor imposing dire financial circumstances upon the Contractor. The Contractor must also utilise every model modern method to continuously record the unforeseeable event both visually by Camera and in writing.

4.12.4 Delay and/or Cost

> If and to the extent that the Contractor suffers delay and/or incurs Cost due to these physical conditions, having complied with Sub-Clauses 4.12.1 to 4.12.3 above, the Contractor shall be entitled subject to Sub-Clause 20.2 [*Claims For Payment and/or EOT*] to EOT and/or payment of such Cost.

The only amendment to this Sub-Clause should be made in the second line for the avoidance of doubt as follows:

> *"... Having complied Sub-Clauses 4.12.1, through 4.12.3 above,"*

The American usage of the word "through" when addressed to periods of time is more effective than the use of the preposition in the English language "to". That is the reason for the proposed amendment.

4.12.5 Agreement or determination of Delay and/or Cost

The agreement or determination, under Sub-Clause 20.2.5 [*Agreement or determination of the Claim*], of any Claim under Sub-Clause 4.12.4 [*Delay and/or Cost*] shall include consideration of whether and (if so) to what extent the physical conditions were Unforeseeable.

It is difficult to see the relevance of this paragraph within Sub-Clause 4.12.5. The entire point of this Sub-Clause is to address physical Conditions, which were unforeseen. If the physical Conditions were unforeseeable then there is the Claim. If the physical Conditions were foreseeable then there is not a Claim.
It might be more appropriate to draft 4.12.5 as follows:

> *"As the agreement or determination, under Sub-Clause 20.2.5 [Agreement and Determination of the Claim], of any Claim under Sub-Clause 4.12.4 [Delay and/or Cost] will proceed in this instance on the basis that there are unforeseeable physical Conditions then the Engineer will take account of any related circumstances as described below."*

The Engineer may also review whether other physical conditions in similar parts of the Works (if any) were more favourable than could reasonably have been foreseen by the Base Date. If and to the extent that these more favourable conditions were encountered, the Engineer may take account of the reductions in Cost, which were due to these conditions in calculating the additional Cost to be agreed or determined under this Sub-Clause 4.12.5. However, the net effect of all additions and reductions under this Sub-Clause 4.12.5 shall not result in a net reduction in the Contract Price.

This provision has now become standard. It encourages the Engineer to carry out a balancing exercise where an unforeseeable physical condition is encountered on a Site where there were variety of physical Conditions all of which anticipated some of which were less onerous in reality than the Contract allowed for.

The Engineer may take account of any evidence of the physical conditions foreseen by the Contractor by the Base Date, which the Contractor may include in the supporting particulars for the Claim under Sub-Clause 20.2.4 [*Fully detailed Claim*], but shall not be bound by any such evidence.

While the Engineer may not be bound by such evidence is also very important that the Engineer shall not wilfully and deliberately discount the evidence.

4.13 Rights of Way and Facilities

The Contractor shall bear all costs and charges for special and/or temporary rights-of-way, which may be required for the purposes of the Works, including those for access to the Site.

The Contractor shall also obtain, at the Contractor's risk and cost, any additional facilities outside the Site, which may be required for the purposes of the Works.

This Sub-Clause is very potentially dangerous and onerous for the Contractor. Initially the Contractor is required to understand and investigate all the necessary Requirements in order to execute the Works effectively. There is no requirement placed upon the Employer to cooperate in carrying out the duties imposed by this Sub-Clause. It could frequently occur that the Contractor will be charged ransom or excessive costs in order to obtain rights of way. There is really no reason the Employer should not be required to provide for the special and/or temporary rights-of-way for access to the Site.

As to the additional facilities outside the Site again the ideal situation would be for there to be any duty upon the Employer to cooperate in assisting the Contractor to obtain these additional facilities. Such facilities may include areas where steel can be prefabricated, Materials stored and concrete batched. In addition, offsite areas may be required to accommodate the Contractor's Personnel.

Any Contractor properly advised should attempt to negotiate the removal of this Sub-Clause prior to signing any Contract.

4.14 Avoidance of Interference

The Contractor shall not interfere unnecessarily or improperly with:

(a) the convenience of the public; or
(b) the access to and use and occupation of all roads and footpaths, irrespective of whether they are public or in the possession of the Employer or of others.

The emphasis here is placed upon the adverbs unnecessarily or improperly. Therefore, where the Contractor exercises due skill and diligence in planning the phased use of the roads and footpath then the impact of this Sub-Clause will be avoided.

> The Contractor shall indemnify and hold the Employer harmless against and from all damages, losses and expenses (including legal fees and expenses) resulting from any such unnecessary or improper interference.

That Second direction within this Sub-Clause is reasonable. The recommendation made earlier in consideration of these Conditions of Contract is that within the brackets the phrase "and expenses" should be removed.

4.15 Access Route

> The Contractor shall be deemed to have been satisfied, at the Base Date, as to the suitability and availability of the access routes to the Site. The Contractor shall take all necessary measures to prevent any road or bridge from being damaged by the Contractor's traffic or by the Contractor's Personnel. These measures shall include the proper use of appropriate vehicles (conforming to legal load and width limits (if any) and any other restrictions) and routes.

Noting the emphasis on Sub-Clause 4.13 as to responsibility for obtaining the access rights this Sub-Clause, 4.15 [*Access Route*], is key for the success or failure of the execution of the Works. It is important to note that this Sub-Clause does not refer to access routes to the facilities necessary for the effective execution of the Site. This Sub-Clause, it is argued, applies equally to such facilities, **but the absence of a clear provision in this sense will give rise to claims.**

It is very important for the Contractor to be aware of any climatic Conditions which may have an inhibiting effect on the use of any roads during certain periods of the year.

> Except as otherwise stated in these Conditions:
>
> (a) the Contractor shall (as between the Parties) be responsible for repair of any damage caused to, and any maintenance which may be required for the Contractor's use of, access routes;

(b) the Contractor shall provide all necessary signs or directions along access routes, and shall obtain any permissions or permits which may be required from the relevant authorities, for the Contractor's use of routes, signs and directions;
(c) the Employer shall not be responsible for any third party claims which may arise from the Contractor's use or otherwise of any access route;
(d) the Employer does not guarantee the suitability or availability of particular access routes; and
(e) all Costs due to non-suitability or non-availability, for the use required by the Contractor, of access routes shall be borne by the Contractor.

The heading to this second direction within this Sub-Clause makes clear that these five directions to the Contractor may well be subject to negotiation as between the Parties. Any Contractor properly advised will attempt to ensure that his use of access routes and all concomitant activities related to access routes will be where possible an exercise of cooperation with the Employer, seeking where possible that the Employer cooperates to ensure the availability of access routes where feasible and practical.

To the extent that non-suitability or non-availability of an access route arises as a result of changes to that access route by the Employer or a third party after the Base Date, and as a result the Contractor suffers delay and/or incurs Cost, the Contractor shall be entitled subject to Sub-Clause 20.2 [*Claims For Payment and/or EOT*] to EOT and/or payment of such Cost.

This is the reasonable concession by the drafters of the Yellow Book and is in addition to the 1999 Yellow Book version.

4.16 Transport of Goods

The Contractor shall:

(a) give a Notice to the Engineer not less than 21 days before the date on which any Plant, or a major item of other Goods (as specified in the Employer's Requirements), will be delivered to the Site;

(b) be responsible for packing, loading, transporting, receiving, unloading, storing and protecting all Goods and other things required for the Works;

(c) be responsible for customs clearance, permits, fees and charges related to the import, transport and handling of all Goods, including all obligations necessary for their delivery to the Site; and

(d) indemnify and hold the Employer harmless against and from all damages, losses and expenses (including legal fees and expenses) resulting from the import, transport and handling of all Goods, and shall negotiate and pay all third party claims arising from their import, transport and handling.

The revisions to the 1999 Yellow Book include the direction stated in sub-paragraph (c) being the responsibility for arranging customs clearance et cetera. This is a logical addition to the other obligations relative to transportation of goods.

It is anticipated that the Contractor, properly advised, would ensure that the Contract with the Parties supplying the particular good from a different country had imposed upon those Parties the replication of the need to indemnify and hold the Employer harmless from anything resulting from a failure to carry out effective packing loading and transporting the particular Goods from the country of supply or manufacture to country containing the Works.

4.17 Contractor's Equipment

The Contractor shall be responsible for all Contractor's Equipment. When brought on to the Site, Contractor's Equipment shall be deemed to be exclusively intended for the execution of the Works. The Contractor shall not remove from the Site any major items of Contractor's Equipment without the Engineer's consent. However, consent shall not be required for vehicles transporting Goods or Contractor's Personnel off Site.

In addition to any Notice given under Sub-Clause 4.16 [*Transport of Goods*], the Contractor shall give a Notice to the Engineer of the date on which any major item of Contractor's

Equipment has been delivered to the Site. This Notice shall be given within 7 days of the delivery date, shall identify whether the item of Contractor's Equipment is owned by the Contractor or a Subcontractor or another person and, if rented or leased, shall identify the rental or leasing entity.

There is no reference in this Sub-Clause to a schedule of Contractor's equipment listed in the Employer's Requirements. Nor is there any reference to any requirement placed upon the Contractor to supply a schedule of equipment to the Engineer within a certain period of time after the issuance of the Letter of Acceptance. There are no periods of time given within which the Contractor shall apply for the Engineer's consent to remove any major item of equipment from the Site. Is such consent required when the major item of equipment is removed for repair?

What is the situation where the Engineer does not consent within a certain period of days? Is the Engineer deemed to have refused consent if within a certain period of days' consent is not given?

These matters are open to debate in the first paragraph of this Sub-Clause.

In the second paragraph a reference is given to a Notice being given within seven days of the delivery date. It is assumed that this Notice is required to be given after the item has been delivered as opposed to anticipating the delivery of the item to Site. Again addition of a few words to this second sentence of the second paragraph would avoid any question such as this.

The direction that the Notice identify the owner of the item of equipment is problematic in the event that the item is leased or rented. The Contractor may not be in a position to identify who is the original owner in the event of a rental or lease agreement. In a revision of this Sub-Clause Word in the last sentence should be removed "and" replaced by "or". Thus, the burden of identifying the owner in the event of a rental will be avoided.

4.18 Protection of the Environment

The Contractor shall take all necessary measures to;

(a) protect the environment (both on and off the Site);
(b) comply with the environmental impact statement for the Works (if any); and
(c) limit damage and nuisance to people and property resulting from pollution, noise and other results of the Contractor's operations and/or activities.

The second obligation is straightforward with regard to compliance if the statement exists.

> The Contractor shall ensure that emissions, surface discharges, effluent and any other pollutants from the Contractor's activities shall exceed neither the values indicated in the Employer's Requirements nor those prescribed by applicable Law.

This obligation is very important for the benefit of the environment in all aspects. And it can be more appropriately enforced than the first three obligations identified in this Sub-Clause. However, this Sub-Clause should be amended follows:

> *"The Contractor shall ensure that all emissions, surface and subsurface discharge, effluent and other pollutants from the Contractor's activities shall not exceed either, where appropriate, the values indicated in the Employer's Requirements, If any, or those values described within applicable Laws regulations et cetera. In the event of a discrepancy between the values the lower value shall apply."*

These amendments take account of situations where the Employer's Requirements are silent and further where the values indicated in the Employer's Requirements, if any, are different from those described in the applicable Laws.

4.19 Temporary Utilities

> The Contractor shall, except as stated below, be responsible for the provision of all temporary utilities, including electricity, gas, telecommunications, water and any other services the Contractor may require for the execution of the Works.

As has been mentioned elsewhere in these Conditions there is no reference here to cooperation between the Employer and the Contractor in situations where it would be totally rational for the Employer to work with the Contractor to ensure the effective provision of all temporary utilities.

Again, it is recommended that the Contractor properly advised will, at the outset, insert within the cooperation element of the Contract reference to the assistance by the Employer in obtaining provision of temporary utilities.

While it is accepted that the subsequent paragraph of this Sub-Clause makes allowance for a situation where the Employer is providing the temporary utilities that situation is not the same as where the onus is on the Contractor to obtain the Utilities and where the cooperation from the Employer would be a great assistance.

> The following provisions of this Sub-Clause shall only apply if, as stated in the Employer's Requirements, the Employer is to provide utilities for the Contractor's use. The Contractor shall be entitled to use, for the purposes of the Works, the utilities on the Site for which details and prices are given in the Employer's Requirements. The Contractor shall, at the Contractor's risk and cost, provide any apparatus necessary for the Contractor's use of these services and for measuring the quantities consumed.
>
> The apparatus provided for measuring quantities consumed shall be subject to the Engineer's consent. The quantities consumed (if any) during each period of payment stated in the Contract Data (if not stated, each month) shall be measured by the Contractor, and the amount to be paid by the Contractor for such quantities (at the prices stated in the Employer's Requirements) shall be included in the relevant Statement.

The apparatus provided for measuring the quantities must be approved by the Engineer. There is no procedure inserted in this Sub-Clause for the approval process. Any revision of this Sub-Clause should include a simple anticipatory procedure for approval or rejection of the apparatus.

4.20 Progress Reports

> Monthly progress reports, in the format stated in the Employer's Requirements (if not stated, in a format acceptable to the Engineer), shall be prepared by the Contractor and submitted to the Engineer. Each progress report shall be submitted in one paper-original, one electronic copy and additional paper copies (if any) as stated in the Contract Data. The first report shall cover the period up to the end of the first month following the Commencement Date. Reports shall be submitted monthly thereafter, each within 7 days after the last day of the month to which it relates.

Yet again there is no procedure provided for the submission of the draft monthly progress report format for approval to the Engineer. It is required that a paper original be submitted. In any revision that this requirement for a paper original become discretionary. There is a reference to an electronic copy. In fact, it is likely that the electronic version will be the original. And from that original will be printed the paper version. This direction should be modified to allow for the realities of the electronic age.

Reporting shall continue until the Date of Completion of the Works or, if outstanding work is listed in the Taking-Over Certificate, the date on which such outstanding work is completed.

There is in fact no reason why reports could not be issued during the DNP.

Unless otherwise stated in the Employer's Requirements, each progress report shall include:

(a) charts, diagrams and detailed descriptions of progress, including each stage of design, Contractor's Documents, procurement, manufacture, delivery to Site, construction, erection, testing, commissioning and trial operation;
(b) photographs and/or video recordings showing the status of manufacture and of progress on and off the Site;
(c) for the manufacture of each main item of Plant and Materials, the name of the manufacturer, manufacture location, percentage progress, and the actual or expected dates of:
    (i) commencement of manufacture,
    (ii) Contractor's inspections,
    (iii) tests, and
    (iv) shipment and arrival at the Site;
(d) the details described in Sub-Clause 6.10 [*Contractor's Records*];
(e) copies of quality management documents, inspection reports, test results, and compliance verification documentation (including certificates of Materials);
(f) a list of Variations, and any Notices given (by either Party) under Sub-Clause 20.2.1 [*Notice of Claim*];
(g) health and safety statistics, including details of any hazardous incidents and activities relating to environmental aspects and public relations; and
(h) comparisons of actual and planned progress, with details of any events or circumstances which may adversely

affect the completion of the Works in accordance with the Programme and the Time for Completion, and the measures being (or to be) adopted to overcome delays.

Item (b) should be amended to state:

"the manufacture of every item of Plant described in the Employer's Requirements and/or for the use of the Contractor in the execution of the Works on Site."

Item (e) should be amended to state:

"Copies of QM Documents, inspection reports, test results and, where applicable, compliance verification documentation including certificates as to the quality of Materials, when appropriate."

Item (f) should be amended to state:

"A list of variations, and any Notices given by either Party under any Sub-Clause of these general Conditions including but not exclusively Sub-Clause 20.2.1 Notice of Claim;"

It is important that every progress report should include every Notice that has been issued by either Party, not merely the Notices relevant to Claims.

Item (g) should be amended to exclude any reference to public relations.

It is appropriate to list any item of public relations in the progress report but not under the heading of health and safety. If anything it should have a separate heading.

Item (h) is very important as it requires the Contractor to consider every event or circumstance and how each of these events or circumstance might adversely affect the progress of the Works.

The Contractor should be pessimistic not optimistic in his view; should the event or circumstances not have an adverse effect then at least allowance has been made but not required.

However, nothing in any progress report shall constitute a Notice under a Sub-Clause of these Conditions.

4.21 Security of the Site

The Contractor shall be responsible for the security of the Site, and:

(a) for keeping unauthorised persons off the Site; and
(b) authorised persons shall be limited to the Contractor's Personnel, the Employer's Personnel, and to any other

personnel identified as authorised personnel (including the Employer's other contractors on the Site), by a Notice from the Employer or the Engineer to the Contractor.

This last direction is awkward.

A preferable direction would be:

"For ensuring that authorised persons are limited to the Contractor's Personnel, the Employer's Personnel, and to all other Personnel identified as authorised, including the Employer's other Contractors, who shall be identified by a Notice from the Employer or the Engineer to the Contractor listing each and every member of the Personnel."

4.22 Contractor's Operations On Site

The Contractor shall confine the Contractor's operations to the Site, and to any additional areas, which may be obtained by the Contractor and acknowledged by the Engineer as working areas. The Contractor shall take all necessary precautions to keep Contractor's Equipment and Contractor's Personnel within the Site and these additional areas, and to keep them off adjacent land.

As far as can be noted this is the first reference to the Engineer acknowledging additional areas as working areas. There therefore needs to be an additional direction to the Contractor here.

"The Contractor if it intends to commence occupation of any additional area shall 14 days before the occupation notify the Engineer as to the size and use of the area occupied. The Engineer shall be deemed to have consented to the use of the area if there is no response to the Notice within seven days of acknowledged receipt."

At all times, the Contractor shall keep the Site free from all unnecessary obstruction, and shall properly store or remove from the Site any Contractor's Equipment (subject to Sub-Clause 4.17 [*Contractor's Equipment*]) and/or surplus materials. The Contractor shall promptly clear away and remove from the Site any wreckage, rubbish, hazardous waste and Temporary Works which are no longer required.

This is a negative instruction. What is required is a positive direction to the Contractor not only to carry out what is set out in this instruction but also to ensure that the Site is tidy and at all times fit for purpose as far as working practices and safety procedures are concerned.

> Promptly after the issue of a Taking-Over Certificate, the Contractor shall clear away and remove, from that part of the Site and Works to which the Taking-Over Certificate refers, all Contractor's Equipment, surplus material, wreckage, rubbish, hazardous waste and Temporary Works. The Contractor shall leave that part of the Site and the Works in a clean and safe condition. However, the Contractor may retain at locations on the Site agreed with the Engineer, during the DNP, such Goods as are required for the Contractor to fulfil obligations under the Contract.

Again, it would be helpful to have a reference to a formal procedure for agreement with the Engineer as to the identity and location of the goods required for the Contractor to fulfil its obligations under the Contract.

4.23 Archaeological and Geological Findings

> All fossils, coins, articles of value or antiquity, and structures and other remains or items of geological or archaeological interest found on the Site shall be placed under the care and authority of the Employer. The Contractor shall take all reasonable precautions to prevent Contractor's Personnel or other persons from removing or damaging any of these findings.

This is frequently a very contentious area for both the Employer and the Contractor as neither of whom wish to be delayed "unnecessarily" by the interventions of archaeological teams.

The practicalities of this instruction and its implementation will depend entirely on the regime contained within the country where the Works are being executed. A democracy will have very different views as to the importance or otherwise of these findings than an oligarchy or a dictatorship.

> The Contractor shall, as soon as practicable after discovery of any such finding, give a Notice to the Engineer in good time to give the Engineer opportunity to promptly inspect and/or

investigate the finding before it is disturbed. This Notice shall describe the finding and the Engineer shall issue instructions for dealing with it.

If the Contractor suffers delay and/or incurs Cost from complying with the Engineer's instructions, the Contractor shall be entitled subject to Sub-Clause 20.2 [*Claims For Payment and/or EOT*] to EOT and/or payment of such Cost.

There is nothing contentious in these provisions.

CHAPTER 5

Clause 5 Design

5.1 General Design Obligations

The Contractor shall carry out, and be responsible for, the design of the Works. Design shall be prepared by designers who:

(a) are engineers or other professionals, qualified, experienced and competent in the disciplines of the design for which they are responsible;
(b) comply with the criteria (if any) stated in the Employer's Requirements; and
(c) are qualified and entitled under applicable Laws to design the Works.

It is important to note that it is not essential for engineers to be the designers of the Works. Other professionals who fall within the qualification criteria set out above are also acceptable as part of the Design team. In the situation where there is a conflict between the qualifications set out in sub-paragraph (a) and those described in the Employer's Requirements, sub-paragraph (b) then a common-sense approach should be taken as to prioritise the qualifications with the main aim being to establish a team which shall provide the capability to Design the Works. If any unresolvable conflict remains between the criteria in sub-paragraph (a) and sub-paragraph (b) then the Contractor should by a Notice seek a determination from the Engineer.

Unless otherwise stated in the Employer's Requirements, the Contractor shall submit to the Engineer for consent the name, address, detailed particulars and relevant experience of each proposed designer/design Subcontractor.

There is no reason for the qualification criteria set out in sub-paragraph (a) to be different from that which shall be provided to the Engineer. Therefore, in any revision on this Clause 5 this direction to the Contractor should be amended to:

> *"Unless otherwise stated in the Employer's Requirements, the Contractor shall submit to the Engineer for consent the name, address, detailed particulars and relevant experience which shall include qualifications and examples of competence in the disciplines of the Design for which they shall be responsible, of each proposed designer/designer Subcontractor."*

The Contractor warrants that the Contractor, the Contractor's designers and design Subcontractors have the experience, capability and competence necessary for the design. The Contractor undertakes that the designers and design Subcontractors shall be available to attend discussions with the Engineer and/or the Employer at all reasonable times (on or off the Site), until the issue of the Performance Certificate.

The first sentence of the Direction in paragraph 1 lacks completeness. It should read:

> *"The Contractor warrants that he, his designers and design Subcontractors have the experience, capability and competency necessary for carrying out and implementing the Design of the Works."*

There is no point in requiring the Contractor to ensure that the entire Design team shall be available throughout the duration of the Design and then the execution of the Works to attend discussions with the Engineer and/or the Employer. What is essential is that there is a core team of designers, whether in the house or subcontract, who are available as required during the execution of the Works.

The alternative would have severe cost implications for the Contractor and is not practical.

Promptly after receiving a Notice under Sub-Clause 8.1 [*Commencement of Works*], the Contractor shall scrutinise the Employer's Requirements (including design criteria and calculations, if any). If the Contractor discovers any error, fault or other defect in the Employer's Requirements, Sub-Clause 1.9 [*Errors in the Employer's Requirements*] shall apply (unless it is an error in the items of reference specified in the Employer's Requirements, in which case Sub-Clause 4.7 [*Setting Out*] shall apply).

It is important to note that Sub-Clause 1.9 [*Errors in the Employer's Requirements*] imposes a continuing burden upon the Contractor to give a Notice to the Engineer whenever the fault, error or defect is discovered. Of course, the key period for perusal of the Employer's Requirements is the 42-day period calculated from the Commencement Date.

In reality these Conditions of Contract do not set out clear procedural timelines for the identification and submission of details for consent to the Engineer with relation to the Commencement Date. The Contractor cannot be expected to appoint his entire Design team until he has been formally notified by the Employer of the acceptance of the Contractor's bid. It is also inevitable that a proportion of the specialist Design team may not be available at the Commencement Date but may only be able to join the team during the Design phase but before the commencement of the execution of the Works. Such a situation reflects reality but should not inhibit the Contractor from notifying the Engineer of the contents of his initial team.

It is important to note that the requirement to issue a Notice to the Engineer applies to each single fault, error or defect and does not intend to await the end of the 42-day period when it might be envisaged that a cumulative Notice could be served.

5.2 Contractor's Documents

> The Contractor's Documents shall comprise the documents:
>
> (a) specified in the Employer's Requirements;
> (b) required to satisfy all permits, permissions, licences and other regulatory approvals which are the Contractor's responsibility under Sub-Clause 1.13 [*Compliance with Laws*]; and
> (c) described in Sub-Clause 5.6 [*As-Built Records*] and Sub-Clause 5.7 [*Operation and Maintenance Manuals*].

An amendment should be made to subparagraph a being:

> *"Specified in the Employer's Requirements as the Contractor's Documents;"*

As to the Documents described in the subsequent sub paragraph it is quite clear that in the event of any doubt as to which Documents are encompassed in those sub-paragraphs then the Contractor should promptly issue a Notice to the Engineer seeking clarification.

5.2.1 Preparation by Contractor

Unless otherwise stated in the Employer's Requirements, the Contractor's Documents shall be written in the language for communications defined in Sub-Clause 1.4 [*Law and Language*].

The Contractor shall prepare all Contractor's Documents, and any other documents necessary to complete and implement the design during execution of the Works and to instruct the Contractor's Personnel. The Employer's Personnel shall have the right to inspect the preparation of all these documents (including any investigation, modelling and testing), wherever they are being prepared.

The order of 5.2.1 is inappropriate. The direction to the Contractor to prepare all Documents et cetera should be in the initial paragraph of this Sub-Clause 5.2.1. Thereafter should follow the direction to the Contractor as to the language in which the Documents shall be written.

5.2.2 Review by Engineer

In this sub-clause 5.2.2

"Review Period" means the period not exceeding 21 days, or as otherwise stated in the Employer's Requirements, calculated from the date on which the Engineer receives a Contractor's Document and a Contractor's Notice;

"Contractor's Document" excludes any of the Contractor's Documents which are not specified in the Employer's Requirements or these Conditions as being required to be submitted for Review, but includes all documents on which a specified Contractor's Document relies for completeness; and

"Contractor's Notice" means the Notice which shall state that the relevant Contractor's Document is considered by the Contractor to be ready for Review under this Sub-Clause 5.2.2 and for use, and that it complies with the Employer's Requirements and these Conditions, or the extent to which it does not do so.

If the Engineer instructs that further Contractor's Documents are reasonably required to demonstrate that the Contractor's design complies with the Contract, the Contractor shall prepare and submit them promptly to the Engineer at the Contractor's cost.

If the Engineer gives a Notice under sub-paragraph (b) above, the Contractor shall:

(i) revise the Contractor's Document;
(ii) resubmit it to the Engineer for Review in accordance with this Sub-Clause 5.2.2, and the Review Period shall be calculated from the date that the Engineer receives it; and
(iii) not be entitled to EOT for any delay caused by any such revision and resubmission and/or by subsequent Review by the Engineer.

If the Employer incurs additional costs as a result of such resubmission and subsequent Review, the Employer shall be entitled subject to Sub-Clause 20.2 [*Claims For Payment and/ or EOT*] to payment by the Contractor of the costs reasonably incurred.

There is every reason why these definitions should be placed in the definitions category. They are not specific to the Design of the Works. In any event the definition of Contractor's Document is unusual. This direction should be rephrased as follows:

> *"Contractor's Documents include all of the Contractor's Documents specified in the Employer's Requirements of these Conditions as being required to be submitted for review and also includes all Documents upon which a specified Contractor's document relies for completeness."*

The negative element of this definition has been removed. The requirement for the Contractor to examine each document in order to test whether it falls within the exclusion is unnecessary.

Further the definition of Contractor's Notice requires amendment:

> *"Contractor's Notice means the Notice which shall formally bring to the Engineers attention the fact that the relevant Contractor's Document is considered by the Contractor to be ready for review under this Clause 5.2.2 and for use."*

The fact that the Contractor has issued this Notice applicable to the relevant Contractor's Document implies that it complies with the Employer's Requirements and these Conditions of Contract and in the situation where the Document does not comply then the extent to which it does not comply and when applicable the reasons for non-compliance.

> If the Employer's Requirements or these Conditions specify that a Contractor's Document is to be submitted to the Engineer for Review, it shall be submitted accordingly, together with a Contractor's Notice.

This is the key direction of Sub-Clause 5.2.2.

However, it fails to include the condition defined in the definition of Contractor's Documents which refers to all Documents upon which a specified Contractor's document relies for completeness. It might be preferable to locate this direction as the first paragraph of the Sub-Clause. Thereafter would follow the definitions. It might be preferable to not omit the introductory words of this key direction in order to assist procedural effectiveness.

Thus:

"Where a Contractor's Document is to be submitted to the Engineer for review it shall be submitted accordingly together with a Contractor's Notice."

This direction avoids the need to refer to the Employer's Requirements and/or the Conditions of Contract. All the Documents, which are relied upon and relevant to the Design and subsequent execution of the Works, must fall within this category of a Contractor's Document.

> The Engineer shall, within the Review Period, give a Notice to the Contractor:
>
> (a) of No-objection (which may include comments concerning minor matters, which will not substantially affect the Works); or
> (b) that the Contractor's Document fails (to the extent stated) to comply, with the Employer's Requirements and/or the Contract, with reasons.

The importance of the Notice of No-objection cannot be sufficiently emphasised. The fact that the Notice can contain comments, instructions, concerning matters, which, in the opinion of the Engineer, will not substantially affect the Works, is fundamental to the progress of the execution of the Design and the Works. Here the Engineer will have made a series of decisions as to whether or not the minor matters will have any substantial effect on the Works. The decision is not that of the Contractor.

Generally and properly in these Conditions of Contract all such burdens are placed upon the Contractor.

The second option given to the Engineer is to issue a Notice identifying the areas where the Contractor's Document fails to comply with the Employer's Requirements and/or the Contract. Strangely the direction as to failure refers to the Employer's Requirements and the Contract, whereas in the definitions for the Contractor's Document or the Contractor's Notice the references are made to the Employer's Requirements or these Conditions. There is no mention made of the Contract. There has to be consistency between these various key directions to the Contractor. The definition of Contractor's Document should be amended as suggested above.

> *"... The Employer's Requirements or these General Conditions which form part of the Contract... ."*

There is no point in referring to General Conditions which do not form part of the Contract and since the guide upon which the Contractor depends is a combination of the Employer's Requirements and the Contract for both the Design and implementation of the Works clarity and consistency are vital. A similar amendment should be made to the definition of Contractor's Notice.

An amendment should be made to the direction to the Engineer where the Document fails as follows:

> *"That the Contractor's Document fails (to the extent stated) to comply with the Employer's Requirements and these General Conditions of Contract as are incorporated in the Contract."*

There can then be no doubt that the Contractor shall only be bound by those General Conditions incorporated into the Contract and at the same direction shall apply to the Engineer when reviewing the Contractor's Documents.

> If the Engineer gives no Notice within the Review Period, the Engineer shall be deemed to have given a Notice of No-objection to the Contractor's Document (provided that all other Contractor's Documents on which that Contractor's Document relies (if any) have been given, or are deemed to have been given, a Notice of No-objection).

It is not possible to give no Notice. This sub-paragraph should be amended as follows:

> *"If the Engineer does not provide any Notice within the Review Period in response to the submission of the Contractor's Document, he shall be deemed to have given a Notice of No-objection to the Contractor's Document, provided that all other Contractor's Documents upon which that Contractor's document relies (if any) have been given, or are deemed to have been given, a Notice of No Objection."*

If the Engineer instructs that further Contractor's Documents are reasonably required to demonstrate that the Contractor's Design compliance with the Contract, the Contractor shall prepare and submit them promptly to the Engineer at the Contractor's cost.

It is difficult to understand why the instruction has to be reasonable. If reasonability is to be taken into account who shall review whether or not the instruction is reasonable? It is the Engineer. What procedure is there for the Contractor to attempt to demonstrate that the instruction is unreasonable? What is to prevent the Employer from seeking additional costs as a result of this request for further Contractor's Documents?

This direction as to requesting additional Contractor's Documents to demonstrate compliance should be deleted. In its place there would be inserted an option (c) as follows:

> *"Where the Contractor's Document fails to demonstrate in an adequate manner that it complies with the Employer's Requirements and/or these General Conditions of Contract and that as a result further Contractor's Documents are required to demonstrate that compliance has taken place."*

This then gives the Engineer the opportunity to seek more information without issuing a Notice stating that the Document does not comply for various reasons. It is a third option.

If the Engineer gives Notice under sub-paragraph (b) above, the Contractor shall:

(i) revise the Contractor's document;
(ii) re-submit it to the Engineer for review in accordance with this Sub-Clause 5.2.2, and the review period shall be calculated from the date that the Engineer receives it; and
(iii) not be entitled to any EOT for any Delay caused by such revision and resubmission and/or by subsequent review by the Engineer.

It is considered that the direction to the Contractor is insufficient. The Engineer has served a Notice under sub paragraph (b) setting out a reasoned response as to why the Document does not comply.

The obvious direction to the Contractor is that:

> *"the Contractor shall revise the Document taking account of each of the reasons upon which the Document failed to comply;"*

Thereafter it is reasonable for the Contractor to re-submit the Document for review.

It is not clear why in these Conditions of Contract it was necessary to state that a Contractor in these circumstances would not be entitled to an EOT. Such an entitlement would only arise should the Contractor find an error, fault or defect in the Employer's Requirements. That opportunity could not arise in the circumstances.

> If the Employer incurs additional costs as a result of such resubmission and subsequent review, he shall be entitled subject to Sub-Clause 20.2 [*Claims for payment and/or EOT*] to payment by the Contractor of the costs reasonably incurred.

It is important to note that the costs referred to here only arise where there has been a necessity for the Engineer to issue a Notice that the Contractor's Document has failed to comply. The implication is clear that the failure to comply must be the fault of the Contractor. Yet there may well be circumstances, not allowed for here, where the failure is a combination of circumstances where both sides carry some liability. The Employer will incur additional cost in those particular circumstances, but it would be unfair to impose those additional costs unilaterally upon the Contractor without taking into account the proportional liability of the Employer.

5.2.3 Construction

> Except for Contractor's Documents under Sub-Clause 5.6 [*As-Built Records*] and Sub-Clause 5.7 [*Operation and Maintenance Manuals*], for each part of the Works requiring Contractor's Documents to be submitted for Review:
>
> (a) construction of such part shall not commence until a Notice of No-objection is given (or is deemed to have been given) by the Engineer for all the Contractor's Documents which are relevant to its design and execution;
> (b) construction of such part shall be in accordance with these Contractor's Documents; and
> (c) the Contractor may modify any design or Contractor's Documents which have previously been submitted for Review, by giving a Notice to the Engineer with reasons. If the Contractor has commenced construction of the part of the Works to which such design or Contractor's Documents are relevant:

(i) work on this part shall be suspended;
(ii) the provisions of Sub-Clause 5.2.2 [*Review by Engineer*] shall apply as if the Engineer had given a Notice in respect of the Contractor's Documents under sub-paragraph (b) of Sub-Clause 5.2.2; and
(iii) work on this part shall not resume until a Notice of No Objection is given (or is deemed to have been given) by the Engineer for the revised documents.

It would have been preferable to refer to the exception separately in this Sub-Clause. Therefore:

> *"For each part of the Works requiring Contractor's Documents to be submitted for review".*

Then a final sub-paragraph shall state:

> *"This Sub-Clause shall not apply to Contractor's Documents as described under Sub-Clause 5.6 [As-built Records] and Sub-Clause 5.7 [Operation and Maintenance Manuals] to which separate reference is made."*

The only change to the wording of the procedures to be followed once the Contractor has submitted a Notice seeking to modify any Design or Contractor's Document is identified in the first direction. In that direction it is stated that work on this part shall be suspended. This direction should be rephrased as follows:

> *"All work on this part shall be immediately suspended;"*

The Contractor should be left in no doubt whatsoever that the suspension is a guillotine process which shall come into effect immediately and it is not a process where by a few days or weeks may pass after the service of the Notice before suspension of work upon the particular part.

5.3 Contractor's Undertaking

The Contractor undertakes that the Design, the Contractor's Documents, the execution of the Works and the completed Works will be in accordance with:

(a) the Laws of the country; and
(b) the Documents forming the Contract, as altered or modified by Variations.

It is strange there is no reference to the Employer's Requirements. There has been a regular reference to the phrase "the Documents forming the Contract". It might be preferable to have a definition in 5.2.2 or at the very beginning in the definitions Clause of the phrase "the Documents forming the Contract".

The first direction is lacking a second verb. The first direction should be as follows:

> *"The Contractor undertakes that the Design, the Contractor's Documents, the execution of the Works and the completed Works will be performed ..."*

The second requirement being that the Undertaking shall be in accordance with the Documents forming the Contract as altered or modified by Variations should be the subject of a condition. As written here the Documents forming the Contract must have been altered or modified by Variations. In fact, it may well be the case that there will not have been any alterations or modifications. If that is the case this direction will not apply. There needs to be a modification as follows:

> *"The Undertaking shall refer to 'the Documents forming the Contract whether altered or modified by variation or not'."*

That qualification allows for any situation.

5.4 Technical Standards and Regulations

> The Contractor's Documents, execution of the Works and the completed work (including defects remedied by the Contractor) shall comply with the Country's technical standards, building, construction and environmental Laws, Laws applicable to the product being produced from the Works, and other standards specified in the Employer's Requirements, applicable to the Works or defined by applicable Laws.

Sub-Clause 5.3 directed the Contractor to give an Undertaking that the Design, the Contractor's Documents, the execution of the Works and the completed Works shall be provided/performed in accordance with the Laws of the country. It seems that Sub-Clause 5.4 is providing a contradiction to Sub-Clause 5.3. The Laws of the country either include all the various technical standards and regulations or they do not. It is far preferable to combine Sub-Clause 5.3 with Sub-Clause 5.4.

There would be a reference to the Laws of the country; thereafter there would be a descriptive Sub-Clause setting out the component element of the Laws and regulations as described in Sub-Clause 5.4.

> All these technical or other standards and Laws shall, in respect of the Works, and each Section and Part, be those in force when the Works or Section or Part are taken under Clause 10 [*Employer's Taking Over*].

It is intended that the Contractor shall be compensated for any changes to the technical or other standards and Laws from the Base Date to the date of the Employer's Taking Over.

Sub-Clause 5.4 continues as follows:

> References in the Contract to published standards shall be understood to be references to the edition applicable on the Base Date, unless stated otherwise. If changed or new applicable standards come into force in the country after the Base Date, the Contractor shall promptly give a Notice to the Engineer and (if appropriate or requested by the Engineer) submit proposals for compliance. To the extent that:
>
> (a) the Engineer considers that that compliance is required and such compliance requires change(s) to the execution of the Works; and
> (b) the Contractor's proposals for compliance constitute a Variation;
>
> then the Engineer shall initiate a Variation in accordance with Clause 13 [*Variations and Adjustments*].

While the intent of this Sub-Clause is helpful, it is not clear. The heading of this Sub-Clause refers to technical standard and regulations. The subtext contained in paragraph 1 refers to a wide range of regulations and Laws, which will be applicable either to the Design, the execution and or the products arising from the Works. The provisions set out in sub-paragraph 3 merely refer to published standards wherever references are made in the Contract. This restriction imposed by means of reference to the Contract and to merely identifying standards is fettering the Contractor.

The whole point of the paragraph is to take into account any change, which occurs from the Base Date until the date where the Employer takes over the Works, in any of the standards, regulations or Laws which are applicable to the

Design, execution of the Works and the product of the Works. In that case all such changes should be the subject of a notification by Notice to the Engineer. All such changes should be accompanied by proposals for compliance whether requested by the Engineer or not.

The third paragraph should be amended as follows:

> *"References in the Contract to published standards shall be understood to be references to the edition published applicable on the Base Date unless stated otherwise. Where a technical or other standard or regulation or law is changed or where a new technical or other standard or regulation or law comes into force in the country after the Base Date, the Contractor shall promptly give a Notice to the Engineer and submit proposals for compliance. To the extent that:*
>
> *(a) the Engineer considers that compliance is required and that such compliance requires change(s) to the execution of the Design, Works or the product of the Works; and*
>
> *(b) the Contractor's proposals for compliance constitute a Variation;*
> *Then the Engineer shall initiate variation in accordance with Clause 13 [Variations and Adjustments]."*

As will be noted, the existing sub-paragraph excludes many of the required options which are available in these circumstances therefore depriving both the Contractor and the Employer of effective benefit of implementing the improved or new standards regulations or Laws which are applicable to the project.

5.5 Training

The Contractor shall carry out training of employees of the Employer (and/or other personnel identified in the Employer's Requirements) in the operation and maintenance of the Works, and any other aspect of the Works, to the extent specified in the Employer's Requirements.

If the Employer's Requirements specify training, which is to be carried out before Taking Over, the Works shall not be considered to be completed for the purposes of Taking Over under Sub-Clause 10.1 [*Taking Over of the Works and Sections*] until this training has been completed in accordance with the Employer's Requirements.

The timing of the training shall be as stated in the Employer's Requirements (if not stated, as acceptable to the Employer). The Contractor shall provide qualified and experienced training staff, training facilities and all training Materials as necessary and/or as stated in the Employer's Requirements.

The first direction is that the Contractor shall give training to the Employer's Personnel in the operation and maintenance of the Works to the extent specified in the Employer's Requirements.

There is no provision here for a situation where there has been an omission in the Employer's Requirements to identify a specific area of training which is as yet essential to the effective operation and maintenance of the Works. If the Employer has not specified anything in his Requirements this will be a Variation and will be paid for. Preferably, an instruction should go into the Guidance Notes on Preparation of Tender Documents.

It is noted that the proposed amendment has deleted reference to "any other aspect of the Works". As no example is given as to what is intended by the drafters of these Conditions of Contract as to what is another aspect it is preferable to remove it.

The timing of the training should be obviously a matter of cooperation between the Employer and the Contractor. The direction to the Contractor to provide qualified and experienced training staff is appropriate. It is also essential that the Contractor provide all necessary training Materials; however whether the Contractor is required to provide training facilities where the Works are in existence appears questionable.

A more appropriate direction to the Contractor might be:

> *"The Contractor shall provide such qualified and experienced training staff as necessary to ensure the effective operation and maintenance of the Works whether stated in the Employer's Requirements or not. The Contractor shall also provide all necessary training Materials for the above purpose. Where the Employer does not have available training facilities then the Contractor will provide all necessary training facilities in order to ensure effective delivery of the training required at the time was Designated by the Employer."*

5.6 As-Built Records

> The Contractor shall prepare, and keep up-to-date, a complete set of "as-built" records of the execution of the Works, showing the exact as-built locations, sizes and details of the work as executed by the Contractor. The format, referencing system, system of electronic storage and other relevant details of the as-built records shall be as stated in the Employer's Requirements (if not stated, as acceptable to the Engineer). These records shall be kept on the Site and shall be used exclusively for the purposes of this Sub-Clause.

The key purpose of these records is to ensure that there is a visual and written record of the progress of the Works. It needs to be clear that the Contractor needs to have a variety of visual aids available and in use 24 hours a day, where appropriate, to ensure that proper and effective records are maintained. The direction that these records are kept on Site is somewhat narrow. It is essential that while records are available on Site that there is a duplicate set available off Site, preferably in an electronic format available to the Contractor, Engineer and the Employer.

The Contractor shall submit to the Engineer under Sub-Clause 5.2.2 [*Review by Engineer*]:

(a) the as-built records for the Works or Section (as the case may be) before the commencement of the Tests on Completion; and
(b) updated as-built records to the extent that any work is executed by the Contractor:
  (i) during and/or after the Tests on Completion, before the issue of any Taking-Over Certificate under Sub-Clause 10.1 [*Taking Over the Works and Sections*]; and
  (ii) after taking over under Sub-Clause 10.1 [*Taking Over the Works and Sections*], before the issue of the Performance Certificate.

In fact, it is preferable for the Contractor to submit the As-built Records regularly, whether monthly or not is a matter of agreement with the Engineer, rather than present the Engineer with a very large and complex file or folder of Records just before the commencement of the Tests on Completion. This Sub-Clause does not reflect the realities of the construction programme and the pressure upon the Engineer.

And amended first direction would be worded as follows:

> *"The Contractor shall submit to the Engineer under Sub-Clause 5.2.2 review by Engineer:*
>
> *(a) The As-built Records for the Works, section or part (as the case may be) regularly, in any event not less than once a month up to the commencement of the Tests on Completion; and..."*

The number of copies of As-built Records to be submitted by the Contractor under this Sub-Clause shall be as required under Sub-Clause 1.8 [*Care and Supply of Documents*].

This is too restrictive. The Engineer will know how many sets of Records he requires. An amendment should suggest:

> *"The number of copies of As-built Records to be submitted by the Contractor under this Sub-Clause shall be as directed by the Engineer or in any event as required under Sub-Clause 1.8 [Care and Supply of Documents]."*

5.7 Operation and Maintenance Manuals

> The Contractor shall prepare, and keep up-to-date, a complete set of operation and maintenance manuals for the Works (the "O&M Manuals" in these Conditions).
>
> The format and other relevant details of the O&M Manuals shall be as stated in the Employer's Requirements and, in any case, these manuals shall:
>
> (a) be in sufficient detail for the Employer to:
>   (i) operate, maintain and adjust the Works to ensure that the performance of the Works, Section and/or Plant (as the case may be) continues to comply with the performance criteria specified in the Employer's Requirements and the Schedule of Performance Guarantees; and
>   (ii) operate, maintain, dismantle, reassemble, adjust and repair the Plant; and
>
> (b) include an inventory of spare parts required for the Employer's future operation and maintenance of the Plant.

There is confusion in the structure of this direction.

The fundamental point of having O&M Manuals is to ensure that the Personnel of the Employer will be able to operate and maintain the Works as originally intended by the Employer.

Therefore, the direction should be as follows:

> *"The format and content of the O&M Manuals shall either be as stated in the Employer's Requirements or as directed by the Engineer and shall, in any event, be sufficiently detailed for professionally qualified Personnel of the Employer to:"*

There is a reference in sub-paragraph (b) to the future operation and maintenance of the Plant. It is considered that when drawn up these Conditions of Contract envisaged that the end result will be a Plant that would produce a product. The Plant will be contained within the Works. Therefore the references in sub-paragraph (a)(i) above to the Works, Section and/or Plant are likely to confuse.

It is possible that the definitions need to be adjusted to take account of the practicalities of the project. It may be that the operation maintenance and adjustment applies to the Plant as opposed to the Works within which the Plant is contained. The inclusion of an inventory of spare parts is helpful but cannot take account of changes whether brought about by innovation or regulation and should not be restrictive.

Before commencement of the Tests on Completion, the Contractor shall submit provisional O&M Manuals for the Works or Section (as the case may be) to the Engineer under Sub-Clause 5.2.2 [*Review by Engineer*].

It would be preferable to give the Engineer a three-month lead-in time to familiarise not only his team but the team of the Employer with the contents of the manuals.

There is no provision in this Sub-Clause for the Employer's Personnel to receive provisional manuals at, or close to, the same time as they are given to the Engineer for review.

These are the Personnel who shall have to maintain and operate and where appropriate adjust the Plant. Allowing the Employer's Personnel access to the provisional manuals prior to the tests taking place ensures that an opportunity is given to the Personnel to comment or review in addition to the comments made by the Engineer. These comments should be not passed directly to the Contractor but be directed to the Engineer.

If during the Tests on Completion any error or defect is found in the provisional O&M Manuals, the Contractor shall promptly rectify the error or defect at the Contractor's risk and cost.

There is no mention of fault in this category of complaints. In other Sub-Clauses there are references to error, fault or defect. Consistency is useful and it is suggested that fault is added to this direction.

Further there is no procedure outlined to notify the Contractor of the error or defect. It will be better for the Employer to list the errors or defects by Notice and publish them to the Engineer for onward action by the Contractor. These manuals are essential for effective operation of the Plant and to ignore a procedure for correction of these provisional manuals is unwise.

Before the issue of any Taking-Over Certificate under Sub-Clause 10.1 [*Taking Over the Works and Sections*], the final O&M Manuals shall be submitted to the Engineer under Sub-Clause 5.2.2 [*Review by Engineer*].

Again it is recommended that the final version of the manuals are submitted to the Employer at the same time as they are submitted to the Engineer. The Employer should then be given a period of time to assimilate the contents of the final version and comment as to the comprehensibility of the version.

5.8 Design Error

> If errors, omissions, ambiguities, inconsistencies, inadequacies or other defects are found in the Contractor's design and/or the Contractor's Documents, they and the Works shall be corrected in accordance with Sub-Clause 7.5 [*Defects and Rejectio*n]. If such Contractor's Documents were previously the subject of a Notice of No-objection given (or deemed to be given) by the Engineer under Sub-Clause 5.2.2 [*Review by Engineer*], the provisions of Sub-Clause 5.2.2 shall apply as if the Engineer had given a Notice in respect of the Contractor's Documents under sub-paragraph (b) of Sub-Clause 5.2.2.
>
> All corrections and resubmissions under this Sub-Clause shall be at the Contractor's risk and cost.

The heading is incorrect. The heading should merely state "design".

The full text of the Sub-Clause refers to not only errors but a variety of other situations. Thus a simple heading referring to design is sufficient. What the Sub-Clause seeks to achieve is a situation where in spite of the fact that the Design (reference omitted in the second sentence) and/or Contractor's Document was the subject of a Notice of No-objection, whether deemed or otherwise, then that Notice of No-objection must be set aside by reason of the discovery of the discrepancy described in the Sub-Clause.

What this Sub-Clause fails to address is the procedure where the design (a very unlikely situation) and/or Contractor's Document has not been submitted to the Engineer under Sub-Clause 5.2.2 review by Engineer.

In that situation: "*the Engineer shall by Notice inform the Contractor of the discrepancy. Thereafter the provisions of Sub-Clause 5.2.2 shall apply as described where the Design or Contractor's Documents had been the subject of a Notice of No-objection.*"

CHAPTER 6

Clause 6 Staff and Labour

6.1 Engagement of Staff and Labour

> Except as otherwise stated in the Employer's Requirements, the Contractor shall make arrangements for the engagement of all Contractor's Personnel, and for their payment, accommodation, feeding, transport and welfare.

This should be read with Sub-Clause 4.8 [*Health and Safety Obligations*]. It should also be read with Sub-Clause 4.1 [*Contractor's General Obligations*] which in paragraph 4 directs the Contractor to be responsible for the safety of all his operations and activities.

6.2 Rates of Wages and Conditions of Labour

> The Contractor shall pay rates of wages, and observe conditions of labour, which comply with all applicable Laws and are not lower than those established for the trade or industry where the work is carried out.
>
> If no established rates or conditions are applicable, the Contractor shall pay rates of wages and observe conditions, which are not lower than the general level of wages and conditions observed locally by employers whose trade or industry is similar to that of the Contractor.

As this direction is intended to be enforceable against the Contractor it might be appropriate to order the Contractor to publish the rates of wages for each of the category of worker employed on the project. There is little seemingly that the Personnel of the Contractor can do against a Contractor who deliberately underpays a certain category of employee other than withdraw their labour by strike action. It may also be that the project envisaged is of such a nature that there is nothing similar in the country that can be used as a comparison whether as to wages or Conditions.

6.3 Recruitment of Persons

> The Contractor shall not recruit, or attempt to recruit, staff and labour from amongst the Employer's Personnel.
>
> Neither the Employer nor the Engineer shall recruit, or attempt to recruit, staff and labour from amongst the Contractor's Personnel.

The heading here is inappropriate. The content of this Sub-Clause is very important. What is intended here is that both sides shall have a strict understanding of the obligation not to attempt to poach the other Parties' Employees. However, this Sub-Clause is rendered useless as there is no sanction imposed for any attempted or actual breach of this obligation.

6.4 Labour Laws

> The Contractor shall comply with all the relevant labour Laws applicable to the Contractor's Personnel, including Laws relating to their employment (including wages and working hours), health, safety, welfare, immigration and emigration, and shall allow them all their legal rights.

Sub-Clause 6.2 addresses rates of wages and Conditions of labour and examines wages in some depth. It appears that Sub-Clause 6.4 under the heading labour laws actually carries out the same exercise without allowing for the flexible approach envisaged under Sub-Clause 6.2.

> The Contractor shall require the Contractor's Personnel to obey all applicable Laws, including those concerning health and safety at work.

There is an earlier reference to the adherence to applicable laws. This obligation is based, very clearly upon the Contractor to ensure that all his Personnel comply with all applicable laws, and the contents of Sub-Clause 4.8 generally. There the Sub-Clause addresses in some detail the health and safety obligations that the Contractor shall ensure are met on Site. The benefits of adding a Sub-Clause 6.4 seems non-existent.

6.5 Working Hours

No work shall be carried out on the Site on locally recognised days of rest, or outside the normal working hours stated in the Contract Data, unless:

(a) otherwise stated in the Contract;
(b) the Engineer gives consent; or
(c) the work is unavoidable or necessary for the protection of life or property or for the safety of the Works, in which case the Contractor shall immediately give a Notice to the Engineer with reasons and describing the work required.

The usual restriction as to working hours is met very quickly by the first option "unless otherwise stated in the Contract".

So that first hurdle is very easily avoided by ensuring that the Contract itself and, where applicable, the Employer's Requirements make clear that the working hours shall be such hours as are reasonable in Works of a similar nature within the country. The second option refers to the consent of the Engineer; it will generally be in the Engineer's interest to see that the Contractor is making progress but obviously not if the labour health and safety and other Requirements are being placed at risk. Although it is not mentioned, having a joint Contractor's workers' council on Site to discuss all such matters and agree them where possible is a very useful entity.

The final exception or possible exception to the stated rule in the first sentence is where the work is unavoidable or necessary for the protection of life or property or for the safety of the Works. In this case the Contractor is required to issue a Notice to the Engineer, which shall contain details of the precise emergency, reasons why out of hours working is required and identifying what Works require to be done where. It is submitted that this analysis of what is required to be given to the Engineer by a Notice is a more accurate description of the various concerns leading to the out of normal hours working.

6.6 Facilities for Staff and Labour

Except as otherwise stated in the Employer's Requirements, the Contractor shall provide and maintain all necessary accommodation and welfare facilities for the Contractor's Personnel.

If such accommodation and facilities are to be located on the Site, except where the Employer has given the Contractor prior permission, they shall be located within the areas identified in the Employer's Requirements. If any such accommodation or facilities are found elsewhere within the Site, the Contractor shall immediately remove them at the Contractor's risk and cost. The Contractor shall also provide facilities for the Employer's Personnel as stated in the Employer's Requirements.

Yet again this is a repetition in slightly different wording of the requirement to provide accommodation and welfare facilities for the Contractor's Personnel. The additional element refers to the locations being within areas identified in the Employer's Requirements.

There is a note to the exception where the Employer has given the Contractor prior permission to use other locations than those identified in the Employer's Requirements. It is noted that there is not required any Notice to be given by the Engineer to the Contractor as to locations found within the Site which are not identified either in the Employer's Requirements or subject to any special exemption.

The Contractor is also obliged to provide like facilities for the Employer's Personnel as stated in the Employer's Requirements. There is no mention here of what is the situation where the Employer's Requirements are silent. It is presumed that in this case no facilities for the Employer's Personnel will be asked for.

6.7 Health and Safety of Personnel

In addition to the requirements of Sub-Clause 4.8 [*Health and Safety Obligations*], the Contractor shall at all times take all necessary precautions to maintain the health and safety of the Contractor's Personnel. In collaboration with local health authorities, the Contractor shall ensure that:

(a) medical staff, first aid facilities, sick bay, ambulance services and any other medical services stated in the Employer's Requirements are available at all times at

> the Site and at any accommodation for Contractor's and Employer's Personnel; and
>
> (b) suitable arrangements are made for all necessary welfare and hygiene requirements and for the prevention of epidemics.

It is difficult to understand the omission of any reference to the Employer's Personnel in this first paragraph. It is thought obvious that under the Sub-Clause addressing cooperation, and the Sub-Clause addressing health and safety obligations, Sub-Clause 4.8, the Contractor should be required at all times to maintain the health and safety of all persons legitimately on the Site. There can be no logical reason to exclude the Employer's Personnel. It is difficult to know what is anticipated in the second direction which addresses necessary welfare and hygiene Requirements. If the intention is to address toilet facilities then that should be addressed clearly. At present the earlier direction addresses first-aid facilities, sickbay and any other medical services stated in the Employer's Requirements.

It is not clear why the same facilities should not be available within any accommodation for the Contractor's and Employer's Personnel. It is preferable to have at least one central point for the sake of keeping proper records of medical accidents and illnesses and also a proper record of treatment given and medication issued. Spreading such facilities around various locations on the Site, depending of course upon the size and practicality of so doing, is not necessarily best practice.

> The Contractor shall appoint a health and safety officer at the Site, responsible for maintaining health, safety and protection against accidents. This officer shall:
>
> (i) be qualified, experienced and competent for this responsibility; and
>
> (ii) have the authority to issue directives for the purpose of maintaining the health and safety of all personnel authorised to enter and/or work on the Site and to take protective measures to prevent accidents.

The concept of appointing a health and safety officer is correct. However, this direction misses a couple of the essential points. The first point is that the Contractor shall appoint such an officer for the Site and such areas of land in addition where Works necessary for the execution of the Works are in progress. The second point is that the officer should be located on Site; the direction is silent as to location.

In addition, it is suggested that should the safety officer be absent for any reason then the Engineer should be notified by Notice and a replacement brought on Site. It is also useful to consider giving the Engineer the power to review the qualifications of the potential safety officer.

The final concern here is that although the officer has authority to issue the directive, there is no indication as to whether they are the equivalent of an instruction issued by the Engineer.

If these directives can be ignored without sanction there is no point in giving such authority to the safety officer.

Throughout the execution of the Works, the Contractor shall provide whatever is required by this person to exercise this responsibility and authority.

This last section of Sub-Clause 6.7 does not meet the concerns raised as to the lack of authority surrounding any directive issued by the safety officer. The Contractor cannot give such authority to the officer; only the Employer or the Engineer can issue instructions/directions. There seems to be a double bind: in that situation, what does the safety officer actually do? He would have some authority within the Contractor's organisation and amongst all those who are put under the Contractor's authority (like Subcontractors, third parties operating within the Contractor's area of authority). What would be the situation with Employer's Personnel when it comes to personal safety issues etc.?

6.8 Contractor's Superintendence

From the Commencement Date until the issue of the Performance Certificate, the Contractor shall provide all necessary superintendence to plan, arrange, direct, manage, inspect, test and monitor the execution of the Works.

Superintendence shall be given by a sufficient number of persons:

(a) who are fluent in or have adequate knowledge of the language for communications (defined in Sub-Clause 1.4 [*Law and Language*]); and
(b) who have adequate knowledge of the operations to be carried out (including the methods and techniques required, the hazards likely to be encountered and methods of preventing accidents),

for the satisfactory and safe execution of the Works.

It is not clear why Sub-Clause 6.8 is separated from Sub-Clause 6.9 which addresses Contractor's Personnel and Key Personnel. It is obvious that the superintendence provided by the Contractor is key to making effective progress for the execution of the Works.

There is no requirement, unlike that of the Contractor's Representative, for the names qualifications and competences of those who are intended to carry out the superintendence to be reviewed by the Engineer. Taking account of the key responsibilities that those providing superintendence must shoulder, it is extraordinary that they will not be reviewed by the Engineer.

6.9 Contractor's Personnel

> The Contractor's Personnel (including Key Personnel, if any) shall be appropriately qualified, skilled, experienced and competent in their respective trades or occupations.

The sanctions for the Contractor failing to provide appropriately skilled and experienced Personnel are obvious and are set out in the second paragraph.

The definition of Key personnel should include at least the Contractor's representative, the Safety officer and the Chief superintendent.

> The Engineer may require the Contractor to remove (or cause to be removed) any person employed on the Site or Works, including the Contractor's Representative and Key Personnel (if any), who:
>
> (a) persists in any misconduct or lack of care;
> (b) carries out duties incompetently or negligently;
> (c) fails to comply with any provision of the Contract;
> (d) persists in any conduct which is prejudicial to safety, health, or the protection of the environment;
> (e) is found, based on reasonable evidence, to have engaged in corrupt, fraudulent, collusive or coercive practice; or
> (f) has been recruited from the Employer's Personnel in breach of Sub-Clause 6.3 [*Recruitment of Persons*].

Most of the actions leading to removal give an opportunity to the member of Personnel to improve his or her conduct. Thus the first cause for removal is where the Personnel has persisted in misconduct or lack of care – that it is a pattern of behaviour that is required to merit a reference to the Engineer and for the

Engineer to take action. The second instance refers to the carrying out of duties incompetently or negligently. Again, that will generally be a pattern of conduct and not a simple single act of negligence or incompetence unless of course the simple act had a fundamental affect upon the progress of the Works.

The third category refers to a situation where the Personnel has failed to comply with any provision of the Contract. There are situations envisaged that occur where a single failure to comply with the provisions of the Contract will result in removal. An example might be where the health and safety provisions have been breached putting the entire workforce at risk. However, usually there will be a pattern of conduct.

The fourth reason for removal is a general sweeping up direction but very important in any event, which refers to conduct, persistent conduct, which is prejudicial to safety, health, or to protection of the environment.

The fifth ground for removal is where a member of the Contractor's Personnel has been found on reasonable evidence to have engaged in various practices which could or would be detrimental to the Contract.

The first question to be raised is what is reasonable evidence? Obviously, a straightforward test is such evidence that on the balance of probability indicates that a corrupt, fraudulent, collusive or coercive practice has been undertaken. However, that useful and generally accepted standard of proof is not present in Sub-Clause 6.9. The next issue concerns the various practices which have been grouped together. The first refers to corruption and it is suggested that it can refer to corruption both on Site and off Site, corruption which led to the awarding the Contract; the scope of possibilities within corruption are vast. Under any revisions the descriptions of practices should be separated so that each may be considered separately and in more detail.

Fraudulent practice is possibly more straightforward as it encompasses attempts to cheat one or other of the Parties by variety of means, whether purely monetary in action or an act which may have a more indirect financial effect where a Contractor's Document may have been forged or an element of the Test on Completion was provided in a manner which allowed the test to succeed when it should have failed. The third element refers to collusive behaviour. It is very difficult to see here, without more, that collusive behaviour could lead to removal from the Site. This practice, collusive behaviour, should be clearly identified with a course of action possibly by referring back to corruption or fraud so that the Parties reading this General Conditions of Contract have some idea what was in the mind of the drafting committee.

Finally, there is a reference to coercive behaviour. This will generally refer to behaviour which is an attempt by a Party to force or persuade the other Party or related Parties to do something that it was unwilling to do. It is not immediately clear what role coercive behaviour plays in Sub-Clause 6.9.

The final cause for removal is where the member of staff has been recruited from the Employer's Personnel in breach of Sub-Clause 6.3. There is no question raised as to Standard of burden of proof or the ease of removal in such circumstances.

If appropriate, the Contractor shall then promptly appoint (or cause to be appointed) a suitable replacement. In the case of the replacement of the Contractor's Representative, Sub-Clause 4.3 [*Contractor's Representative*] shall apply. In the case of the replacement of Key Personnel (if any), Sub-Clause 6.12 [*Key Personnel*] shall apply.

The procedures for appointment, where appropriate, are straightforward.

6.10 Contractor's Records

Unless otherwise proposed by the Contractor and agreed by the Engineer, in each progress report under Sub-Clause 4.20 [*Progress Reports*], the Contractor shall include records of:

(a) occupations and actual working hours of each class of Contractor's Personnel;
(b) the type and actual working hours of each of the Contractor's Equipment;
(c) the types of Temporary Works used;
(d) the types of Plant installed in the Permanent Works; and
(e) the quantities and types of Materials used for each work activity shown in the Programme, at each work location and for each day of work.

The heading here is "Contractor's Records". However, there is no mention of a specific document headed "Contractor's Records". **The definition of Contractor's Documents should be amended accordingly.**

This initial direction to the Contractor should be amended as follows:

> *"As part of the Contractor's Documents there shall be attached Documents known as Contractor's Records. These daily records shall subsequently form part of the progress reports required under Sub-Clause 4.20 [Progress Reports] and the Contractor shall include the Records of:"*

An additional amendment might be made to the second item in the schedule being the type of actual working hours for each of the Contractor's Equipment. With certain types of Contractor's Equipment there are Personnel allotted to the specific item of equipment and it might well be that it is essential to record the hours for each man attending the particular item of equipment.

To require the Contractor to follow a simple reporting format which allocates the hours and times of a piece of Equipment to Plant and all Materials used is essential for the effective management of the Contract and further should the question of daywork ever arise when the Contractor can simply and easily identify which of the various items of equipment and various trades and occupations should be allocated to the Daywork.

6.11 Disorderly Conduct

The Contractor shall at all times take all necessary precautions to prevent any unlawful, riotous or disorderly conduct by or amongst the Contractor's Personnel, and to preserve peace and protection of persons and property on and near the Site.

It is omitted here that the direction for prevention and preservation apply where the Personnel concerned are other Parties or persons who are legally authorised to be on Site. It would be unfair and unreasonable to impose such a strict obligation on the Contractor in the event that there were illegal and/or unauthorised persons on the Site or close to the Site causing disorderly conduct.

6.12 Key Personnel

If no Key Personnel are specified in the Employer's Requirements this Sub-Clause shall not apply.

The Contractor shall appoint the natural persons named in the Tender to the positions of Key Personnel. If not so named, or if an appointed person fails to act in the relevant position of Key Personnel, the Contractor shall submit to the Engineer for consent the name and particulars of another person the Contractor proposes to appoint to such position. If consent is withheld or subsequently revoked, the Contractor shall similarly submit the name and particulars of a suitable replacement for such position.

If the Engineer does not respond within 14 days after receiving any such submission, by giving a Notice stating his/her objection to the appointment of such person (or replacement) with reasons, the Engineer shall be deemed to have given his/her consent.

The Contractor shall not, without the Engineer's prior consent, revoke the appointment of any of the Key Personnel or appoint a replacement (unless the person is unable to act as a result of death, illness, disability or resignation, in which case the appointment shall be deemed to have been revoked with immediate effect and the appointment of a replacement shall be treated as a temporary appointment until the Engineer gives his/her consent to this replacement, or another replacement is appointed, under this Sub-Clause).

All Key Personnel shall be based at the Site (or, where Works are being executed off the Site, at the location of the Works) for the whole time that the Works are being executed. If any of the Key Personnel is to be temporarily absent during execution of the Works, a suitable replacement shall be temporarily appointed, subject to the Engineer's prior consent.

All Key Personnel shall be fluent in the language for communications defined in Sub-Clause 1.4 [*Law and Language*].

On the assumption that this Sub-Clause applies, a few points are raised. **Is the Contractor's Representative a Key Personnel or not? If he is, then this Sub-Clause always applies, irrespective of instructions in the Employer's Requirements.** In the first paragraph the Contractor is directed to appoint the natural persons named in the tender. What is intended here is that there should be a schedule of identified individuals proposed for the posts contained under the heading Key Personnel. There is no requirement to identify the roles of natural persons. Where the Employer's Requirements identify roles but not specific named individuals then there is a procedure for submission of names and appropriate roles for consent or objection. However, in the first paragraph, there is a reference to the situation where the appointed person, presumably the named person, fails to act.

There is no indication given within the Sub-Clause of examples of failure to act. In paragraph 4 of this Sub-Clause there are situations described which address where the person is unable to act. This is not a failure to act. This is a particular inability and the examples given support that analysis. Examples include an inability to act as a result of death, disability or resignation et cetera.

Therefore, it is clear that while the Sub-Clause identified a situation where the person has failed to act, the Sub-Clause had failed to give supporting examples of situations identifying such failures.

CHAPTER 7

Clause 7 Plant, Materials and Workmanship

7.1 Manner of Execution

The Contractor shall carry out the manufacture, supply, installation, testing and commissioning and/or repair of Plant, the production, manufacture, supply and testing of Materials, and all other operations and activities during the execution of the Works:

(a) in the manner (if any) specified in the Contract;
(b) in a proper workmanlike and careful manner, in accordance with recognised good practice; and
(c) with properly equipped facilities and non-hazardous Materials, except as otherwise specified in the Contract.

This definition emphasises the execution of the Plant and the Materials. This Clause addresses workmanship in general. Therefore, Sub-Clause 7.1 is intended to address workmanship in the execution of the Works. Sub-Clause 7.1 does not appear to address the Works as such. The focus of Sub-Clause 7.1 is on Plant and Materials and all other operations and activities during execution of the Works. The first paragraph of Sub-Clause 7.1 should be amended as follows:

> *"The Contractor shall carry out the execution of the Work including manufacture, supply, installation, testing and commissioning and/or repair of Plant, reproduction, manufacture, supply and testing of Materials:"*

At present there is a reference to all other operations and activities during the execution of the Works but there is no focus upon the Works themselves. The execution of the Works shall be carried out in a proper workmanlike and careful manner in accordance with recognised good practice. This is very subjective and could lead to conflict, as what is recognised good practice in one country is not necessarily so recognised in another. Therefore, it might be a useful addition to add into the second direction:

> *"In a proper workman like and careful manner, in accordance with the recognised good practice within the country in which the Works are being executed; and ..."*

However, with respect to direction in sub-paragraph (c) which requires the manner of execution to be carried out with properly equipped facilities and non-hazardous Materials the previous direction added reference to the country in which the Works are being carried out whereas here there is a reference to international practice when identifying properly equipped facilities. There should be consistency in these directions.

7.2 Samples

The Contractor shall submit the following samples of Materials, and relevant information, to the Engineer for consent prior to using the Materials in or for the Works:

(a) manufacturer's standard samples of Materials and samples specified in the Contract, all at the Contractor's cost; and
(b) additional samples instructed by the Engineer as a Variation.

The concern here relates to Materials that are identified in the Employer's Requirements but not specified in the Contract. If common sense is to be used because the Materials and samples are not specified in the Contract all the Parties may refer to the Employer's Requirements and act accordingly. Further it appears that the intent is that where it is intended to use a variety of Materials, standard samples of those Materials shall be supplied to the Engineer for review under Sub-Clause 5.2.2. Then what is the situation? Continuing submission of the individual samples seems sensible.

The referencing direction of this Sub-Clause to standard samples of Materials and sample specified in the Contract appear to refer to disparate samples. In other words, there may be samples of sand, gravel, cement et cetera which may be deemed manufacturer's standard samples of Materials and on the other hand there may be specific additives added to the concrete to strengthen it for certain uses that will not be standard and that will have to be identified and supplied by the Contractor. This direction should be is reformulated as follows:

> *"Manufacturers standard samples of Materials and where specified in the Contract such additional samples as the Employer wishes to identify."*

Each sample shall be labelled as to origin and intended use in the Works.

It is recommended that the changes made to this last direction as follows:

> *"Each sample shall be where appropriate placed in a transparent container which shall contain on its surface information as to the date, Country and name of the supplier of the sample, together with the names and contact details of any intermediary agents through which the sample had passed."*

7.3 Inspection

The Employer's Personnel shall, during all the normal working hours stated in the Contract Data and at all other reasonable times:

(a) have full access to all parts of the Site and to all places from which natural Materials are being obtained;
(b) during production, manufacture and construction (at the Site and elsewhere), be entitled to:
 (i) examine, inspect, measure and test (to the extent stated in the Employer's Requirements) the Materials, Plant and workmanship,
 (ii) check the progress of manufacture of Plant and production and manufacture of Materials, and
 (iii) make records (including photographs and/or video recordings); and
(c) carry out other duties and inspections, as specified in these Conditions and the Employer's Requirements.

The Contractor shall give the Employer's Personnel full opportunity to carry out these activities, including providing safe access, facilities, permissions and safety equipment.

The Contractor shall give a Notice to the Engineer whenever any Materials, Plant or work is ready for inspection, and before it is to be covered up, put out of sight, or packaged for storage or transport. The Employer's Personnel shall then either carry out the examination, inspection, measurement or testing without unreasonable delay, or the Engineer shall promptly give a Notice to the Contractor that the Employer's Personnel do not require to do so. If the Engineer gives no such Notice and/or the Employer's Personnel do not attend at the time stated in the Contractor's Notice (or such time as may be agreed with the Contractor), the Contractor may proceed with covering up, putting out of sight or packaging for storage or transport.

If the Contractor fails to give a Notice in accordance with this Sub-Clause, the Contractor shall, if and when required by the Engineer, uncover the work and thereafter reinstate and make good, all at the Contractor's risk and cost.

This Clause addresses Plant, Materials and Workmanship. Therefore, this direction in Sub-Clause 7.3 refers not only to Materials and Plant but also to the day-to-day execution of the Works. That is not clear from the first line of the last paragraph of this Sub-Clause.

This last paragraph should be amended as follows:

> *"The Contractor shall give a Notice to the Engineer whenever any Materials, Plant or element of the Works is ready for inspection, before it is covered up, put out of sight, or packaged for storage or transport."*

The word "and" has been removed as it confuses the issue. Focus has been redirected upon not only Materials and Plant but also in the day-to-day progress of the Works.

It is essential that the Engineer is given a clear warning by Notice whenever a portion of the Works is about to be constructed, therefore covered up, so that the Engineer can inspect the element of the Site prior to the Works commencing. The emphasis, properly given, that should the Employer's Personnel not attend at the time stated in the Contractor's Notice then the Contractor may proceed with the Works is to be treated with great caution.

The first line of this paragraph does not require the Contractor to insert within the Notice a time when the inspection should be carried out. As this is key to the situation where the Employer's Personnel have not attended at the time stated, the first sentence should be amended as follows:

> *"The Contractor shall give a timely Notice to the Engineer whenever any Material, Plant or element of the Works is ready for inspection stating a time when inspection should take place before it is covered up, put out of sight or packaged for storage or transport."*

7.4 Testing by the Contractor

This Sub-Clause shall apply to all tests specified in the Contract, other than the Tests after Completion (if any).

The Contractor shall provide all apparatus, assistance, documents and other information, temporary supplies of electricity and water, equipment, fuel, consumables, instruments, labour, materials, and suitably qualified, experienced and competent staff, as are necessary to carry out the specified tests efficiently and properly. All apparatus, equipment and instruments shall be calibrated in accordance with the standards specified in the Employer's Requirements or defined by applicable Laws and, if requested by the Engineer, the Contractor shall submit calibration certificates before carrying out testing.

The Contractor shall give a Notice to the Engineer, stating the time and place for the specified testing of any Plant, Materials and other parts of the Works. This Notice shall be given in reasonable time, having regard to the location of the testing, for the Employer's Personnel to attend.

The Engineer may, under Clause 13 [*Variations and Adjustments*], vary the location or timing or details of specified tests, or instruct the Contractor to carry out additional tests. If these varied or additional tests show that the tested Plant, Materials or workmanship is not in accordance with the Contract, the Cost and any delay incurred in carrying out this Variation shall be borne by the Contractor.

This direction within paragraph 4 of this Sub-Clause does not make sense. The first sentence authorises the Engineer by reason of Clause 13 to vary the location or timing or details of the specified tests. No reason is given as to why the Engineer may exercise this power as a result of operation of Clause 13. It is sensible to give the Engineer discretion to vary the location and timing of the tests, but it is not clear why the Engineer should have the authority to vary the details of the tests.

Secondly this first direction in paragraph 4 authorises the Engineer to instruct the Contractor to carry out additional tests. This ability to require additional tests is not linked to Sub-Clause 7.5 [*Defects and Rejection*]. That Sub-Clause sets out a very detailed procedure in the event of the test having a defect or being rejected. It is quite unclear in the event that the test does not fail or is not rejected what is the purpose of carrying out additional tests.

This paragraph should be redrafted:

> *"The Engineer may, under Clause 13 [Variation and Adjustments] vary the location or timing or details of the specified tests. Further the Engineer may, if the Engineer considers the instruction appropriate, instruct the Contractor to carry out additional tests. If these varied tests show that the tested Plant, Materials or workmanship is not in accordance with the Employer's Requirements, the cost and any Delay incurred in carrying out this variation shall be borne by the Contractor."*

It is not clear how this paragraph assists the effective management of the Contract when the ability to require the Contractor to submit any material, Plant or element of the execution of the Works to a retest is clearly set out in Sub-Clause 7.5.

The Engineer shall give a Notice to the Contractor of not less than 72 hours of his/her intention to attend the tests. If the Engineer does not attend at the time and place stated in the Contractor's Notice under this Sub-Clause, the Contractor may proceed with the tests, unless otherwise instructed by the Engineer, which tests shall then be deemed to have been made in the Engineer's presence. If the Contractor suffers delay and/or incurs Cost from complying with any such instruction or as a result of a delay for which the Employer is responsible, the Contractor shall be entitled subject to Sub-Clause 20.2 [*Claims For Payment and/or EOT*] to EOT and/or payment of Cost Plus Profit.

This paragraph directs the Engineer to acknowledge his intention to attend by a Notice. There is then followed a reference to the situation where the Engineer does not attend. It is thought that a more effective structure would be:

> *"if the Engineer is unable to attend at the time and place stated in the Contractor's Notice then the Engineer shall issue a Notice instructing the Contractor when attendance can be made."*

The risk placed upon the Contractor of proceeding with the test in the absence of the Engineer without clear instruction is too great. If the Engineer cannot attend as proposed, then an alternative arrangement should be made and if there is unnecessary delay then the cost incurred should be borne by the Employer. An interim stage would then link in with the last element of this paragraph, which provides for a situation where the tests have been delayed or costs incurred as a result of obedience to the Engineer's instruction or Delay on the part of the Employer.

If the Contractor causes any delay to specified tests (including varied or additional tests) and such delay causes the Employer to incur costs, the Employer shall be entitled subject to Sub-Clause 20.2 [*Claims For Payment and/or EOT*] to payment of these costs by the Contractor.

The Contractor shall promptly forward to the Engineer duly certified reports of the tests. When the specified tests have been passed, the Engineer shall endorse the Contractor's test certificate, or issue a test certificate to the Contractor, to that effect. If the Engineer has not attended the tests, he/she shall be deemed to have accepted the readings as accurate.

This direction is not explicit. Yet it contains the key to the testing approval process.

A better structure is as follows:

> *"Once the tests have been completed the Contractor shall promptly forward to the Engineer duly certified reports of the tests.*
>
> *If the Engineer has not attended the test, he shall be deemed to have accepted the reports as accurate whatever the result."*

Sub-Clause 7.5 [*Defects and Rejection*] shall apply in the event that any Plant, Materials and other parts of the Works fails to pass a specified test.

This is self-evident.

7.5 Defects and Rejection

If, as a result of an examination, inspection, measurement or testing, any Plant, Materials, Design or workmanship is found to be defective or otherwise not in accordance with the Contract, the Engineer shall give a Notice to the Contractor describing the item of Plant, Materials, Design or workmanship that has been found to be defective. The Contractor shall then promptly prepare and submit a proposal for necessary remedial work.

The Engineer may Review this proposal, and may give a Notice to the Contractor stating the extent to which the proposed work, if carried out, would not result in the Plant, Materials, Design or workmanship complying with the Contract. After receiving such a Notice the Contractor shall promptly submit a revised proposal to the Engineer. If the Engineer gives no such Notice within 14 days after receiving the Contractor's proposal (or revised proposal), the Engineer shall be deemed to have given a Notice of No-objection.

This is a reasonable definition. It is clear from the first sentence that the heading to Sub-Clause 7.3 requires expansion to include measurement and examination. However, the first paragraph of Sub-Clause 7.5 does not identify the concept of rejection. Therefore, this first paragraph needs amending to include that concept.

> *"If, as a result of an examination, inspection, measurement or testing, any Plant, Materials, Design or workmanship is found to be defective or otherwise not in accordance with the Employer's Requirements and therefore rejected, the Engineer shall give a Notice to the Contractor describing the item of Plant, Materials, Design or workmanship that has been found to be defective and therefore shall be rejected."*

The definition as now written takes note of the change from the Contract to Employer's Requirements and further makes explicit the fact that the defect has resulted in the item being rejected. There will be situations where the defect is minor or not substantial and where there is no need for rejection.

The next direction as to review by the Engineer is awkward. A better definition might be:

> *"The Engineer may review this proposal and subsequently give a Notice to the Contractor of No-objection which may state the extent to which the proposal does not comply with the Employer's Requirements the Engineer may issue a Notice of Rejection identifying the extent to which the proposal does not comply with the Employer's Requirements."*

If the Contractor fails to promptly submit a proposal (or revised proposal) for remedial work, or fails to carry out the proposed remedial work to which the Engineer has given (or is deemed to have given) a Notice of No-objection, the Engineer may:

(a) instruct the Contractor under sub-paragraph (a) and/or (b) of Sub-Clause 7.6 [*Remedial Work*]; or
(b) reject the design, Plant, Materials or workmanship by giving a Notice to the Contractor, with reasons, in which case sub-paragraph (a) of Sub-Clause 11.4 [*Failure to Remedy Defects*] shall apply.

This direction to the Engineer could create concerns. A failure to submit a proposal or revised proposal promptly is reasonable. However, to link that failure with a failure to carry out the proposed Remedial work without a link to promptness is problematic.

A better solution might be:

> *"If the Contractor fails to submit promptly a proposal, or revised proposal where appropriate, for the remedial work then the Engineer may:*
>
> *(a) instruct the Contractor under sub-paragraph (a) and/or (b) of Sub-Clause 7.6 [Remedial Work]; or*

(b) *reject the Design, Plant, Materials or workmanship by giving a Notice to the Contractor, with reasons, after receipt of which the Employer may (as the Employer's discretion) employ and pay other persons to make good the defect and shall be entitled, subject to Sub-Clause 20.2 [Claims for Payment and/or EOT] to payment by the Contractor of all costs arising from the Contractor's Failure.*

"If the Contractor fails to carry out promptly the proposed remedial work to which the Engineer has given out (or is deemed to have given) a Notice of No-objection the Engineer may exercise the options set out in the previous paragraph."

After remedying defects in any Plant, Materials, Design or workmanship, if the Engineer requires any such items to be retested, the tests shall be repeated in accordance with Sub-Clause 7.4 [*Testing by the Contractor*] at the Contractor's risk and cost. If the rejection and retesting cause the Employer to incur additional costs, the Employer shall be entitled subject to Sub-Clause 20.2 [*Claims For Payment and/or EOT*] to payment of these costs by the Contractor.

This direction is rational.

7.6 Remedial Work

In addition to any previous examination, inspection, measurement or testing, or test certificate or Notice of No-objection by the Engineer, at any time before the issue of the Taking-Over Certificate for the Works the Engineer may instruct the Contractor to:

(a) repair or remedy (if necessary, off the Site), or remove from the Site and replace any Plant or Materials which are not in accordance with the Contract;
(b) repair or remedy, or remove and re-execute, any other work which is not in accordance with the Contract; and
(c) carry out any remedial work which is urgently required for the safety of the Works, whether because of an accident, unforeseeable event or otherwise.

The Contractor shall comply with the instruction as soon as practicable and not later than the time (if any) specified in the instruction, or immediately if urgency is specified under sub-paragraph (c) above.

Remedial work

The reference to Contract needs to be changed to Employer's Requirements as it is likely that the criteria are specified there. There is nothing in this first paragraph that suggests how the Engineer may, after having seemingly examined, inspected and approved the Plant, Materials or an element of workmanship, subsequently order (by Notice?) repairs or remedies by reason of non-compliance with the Employer's Requirements. It is wondered how the Engineer is able to have this subsequent authority in the face of a series of approvals or No-objections.

The Contractor shall bear the cost of all remedial work required under this Sub-Clause, except to the extent that any work under sub-paragraph (c) above is attributable to:

(i) any act by the Employer or the Employer's Personnel. If the Contractor suffers delay and/or incurs Cost in carrying out such work, the Contractor shall be entitled subject to Sub-Clause 20.2 [*Claims For Payment and/or EOT*] to EOT and/or payment of such Cost Plus Profit; or

(ii) an Exceptional Event, in which case Sub-Clause 18.4 [*Consequences of an Exceptional Event*] shall apply.

If the Contractor fails to comply with the Engineer's instruction, the Employer may (at the Employer's sole discretion) employ and pay other persons to carry out the work. Except to the extent that the Contractor would have been entitled to payment for work under this Sub-Clause, the Employer shall be entitled subject to Sub-Clause 20.2 [*Claims For Payment and/or EOT*] to payment by the Contractor of all costs arising from this failure. This entitlement shall be without prejudice to any other rights the Employer may have, under the Contract or otherwise.

This element is self-explanatory.

7.7 Ownership of Plant and Materials

Each item of Plant and Materials shall, to the extent consistent with the mandatory requirements of the Laws of the Country, become the property of the Employer at whichever is the earlier of the following times, free from liens and other encumbrances:

(a) when it is delivered to the Site;
(b) when the Contractor is paid the value of the Plant and Materials under Sub-Clause 8.11 [*Payment for Plant and Materials after Employer's Suspension*]; or
(c) when the Contractor is paid the amount determined for the Plant and Materials under Sub-Clause 14.5 [*Plant and Materials intended for the Works*].

Option (a) means that whenever an item of Plant or Material is delivered to the site it becomes property of the Employer, notwithstanding payment or absence thereof?

Option (c) only applies where the Plant or Material is listed in the Contract Data or the Employer's Requirements as being an item for which payment shall be made as soon as the item is shipped. Otherwise this definition is self-explanatory.

7.8 Royalties

Unless otherwise stated in the Employer's Requirements, the Contractor shall pay all royalties, rents and other payments for:

(a) natural Materials obtained from outside the Site; and
(b) the disposal of Material from demolitions and excavations and of other surplus Material (whether natural or man-made), except to the extent that disposal areas within the Site are specified in the Employer's Requirements.

The heading is incomplete. What is intended is that the Contractor shall pay royalties and other fees for natural Materials brought into the Site from the surrounding area and further that in the event of demolition and excavation the Contractor shall pay the costs of transporting and disposing of the detritus or excavated Material. Therefore, the heading needs to be expanded to: Royalties, Disposal Costs.

CHAPTER 8

Clause 8 Commencement, Delays and Suspension

8.1 Commencement of Works

The Engineer shall give a Notice to the Contractor stating the Commencement Date, not less than 14 days before the Commencement Date. Unless otherwise stated in the Particular Conditions, the Commencement Date shall be within 42 days after the Contractor receives the Letter of Acceptance.

The Contractor shall commence the execution of the Works on, or as soon as is reasonably practicable after, the Commencement Date and shall then proceed with the Works with due expedition and without delay.

This definition needs to be amended to take account of alternative documentation. Thus, the first paragraph should read:

> *"The Engineer shall, not less than 14 days before the commencement date, issue a Notice to the Contractor identifying the Commencement Date. Unless otherwise stated in the Contract documents, the Commencement Date shall be within 42 days after the Contractor received the Letter of Acceptance."*

This amended definition allows for a statement as to how to determine the Commencement Date to be found in a variety of alternative Documents. Definitions should not of themselves limit the choice of Documents.

8.2 Time for Completion

The Contractor shall complete the whole of the Works, and each Section (if any), within the Time for Completion for the Works or Section (as the case may be), including completion of all work which is stated in the Contract as being required for the Works or Section to be considered to be completed for the purposes of taking over under Sub-Clause 10.1 [*Taking Over the Works and Sections*].

There are two concerns with this definition. The first concern is that elements of the Works identified as parts are ignored in this definition. It is also noted that "parts" are not included in the final element of this definition.

Secondly there is in the second leg of the test a reference to the "Contract". There may not be a Contract per se.

Therefore, this definition should be amended:

> *"The Contractor shall complete the whole of the Works, each Section (if any) and each Part (if any), within the Time for Completion for the Works, Section or Part (as the case may be), including:*
>
> *(a) achieving the passing of the Tests on Completion; and*
> *(b) completing all work which is stated in the Contract, Employer's Requirements or elsewhere as being required for the Works, Section or Part to be considered to be completed for the purposes of taking over under Sub-Clause 10. 1 [Taking over the Works and Sections] and Sub-Clause 10.2[Taking over Parts]."*

8.3 Programme

The Contractor shall submit an initial programme for the execution of the Works to the Engineer within 28 days after receiving the Notice under Sub-Clause 8.1 [*Commencement of Works*]. This programme shall be prepared using programming software stated in the Employer's Requirements (if not stated, the programming software acceptable to the Engineer). The Contractor shall also submit a revised programme, which accurately reflects the actual progress of the Works, whenever any programme ceases to reflect actual progress or is otherwise inconsistent with the Contractor's obligations.

It is correct to refer to programming software as being compatible with the Employer's Requirements whether explicitly stated or not. However, the need to refer to acceptability of the Engineer without a simple procedure for obtaining that acceptance is not sufficient.

Therefore, this definition needs to be amended to provide a simple procedure.

> *"The Contractor shall submit an initial programme for the execution of the Works to the Engineer within 28 days after receiving the Notice provided under Sub-Clause 8.1 [Commencement of Works]. This programme shall be prepared using programming software stated in the Employer's Requirements or elsewhere in the Contract*

Data; if not stated then within seven days of receipt of the Notice provided under Sub-Clause 8.1[Commencement of Works] the Contractor shall submit to the Engineer for review and approval programming software. The Engineer shall within 7 days of receipt of this proposal either issue a Notice of No Objection or a Notice identifying an alternative programming software acceptable to the Engineer.

The Contractor shall subsequently submit an updated programme at a frequency identified by the Engineer or monthly, which accurately reflects the actual progress of the Works.

Whenever the programme ceases to reflect actual progress or is otherwise inconsistent with the Contractor's obligations then the Contractor shall promptly submit a revised programme which reflects actual progress or the inconsistency."

This definition is now considerably expanded and, it is believed, more accurately reflects the needs of both the Employer and the Contractor in establishing and providing flexibility for the programme.

The initial programme and each revised programme shall be submitted to the Engineer in one paper copy, one electronic copy and additional paper copies (if any) as stated in the Contract Data, and shall include:

(a) the Commencement Date and the Time for Completion, of the Works and of each Section (if any);
(b) the date right of access to and possession of (each part of) the Site is to be given to the Contractor in accordance with the time (or times) stated in the Contract Data. If not so stated, the dates the Contractor requires the Employer to give right of access to and possession of (each part of) the Site;
(c) the order in which the Contractor intends to carry out the Works, including the anticipated timing of each stage of design, preparation and submission of Contractor's Documents, procurement, manufacture, inspection, delivery to Site, construction, erection, installation, work to be undertaken by any Nominated Subcontractor (as defined in Clause 4.5 [*Nominated Subcontractors*]), testing, commissioning and trial operation;
(d) the Review periods under Sub-Clause 5.2.2 [*Review by Engineer*], and periods for Review for any other submissions specified in the Employer's Requirements or required under these Conditions;

(e) the sequence and timing of inspections and tests specified in, or required by, the Contract;
(f) for a revised programme: the sequence and timing of the remedial work (if any) to which the Engineer has given a Notice of No-objection under Sub-Clause 7.5 [*Defects and Rejection*] and/or the remedial work (if any) instructed under Sub-Clause 7.6 [*Remedial Work*];
(g) all activities (to the level of detail specified in the Employer's Requirements), logically linked and showing the earliest and latest start and finish dates for each activity, the float (if any), and the critical path(s);
(h) the dates of all locally recognised days of rest and holiday periods (if any);
(i) all key delivery dates of Plant and Materials;
(j) for a revised programme and for each activity: the actual progress to date, any delay to such progress and the effects of such delay on other activities (if any); and
(k) a supporting report which includes:
  (i) a description of all the major stages of the execution of the Works;
  (ii) a general description of the methods which the Contractor intends to adopt in the execution of the Works;
  (iii) details showing the Contractor's reasonable estimate of the number of each class of Contractor's Personnel, and of each type of Contractor's Equipment, required on the Site, for each major stage of the execution of the Works;
  (iv) if a revised programme, identification of any significant change(s) to the previous programme submitted by the Contractor; and
  (v) the Contractor's proposals to overcome the effects of any delay(s) on progress of the Works.

This is an extremely long definition. There are amendments to be proposed. As the software program is indeed an IT program it appears outdated to refer to a paper copy. The only time a paper copy should be provided is if it is required within the Employer's Requirements or in the Contract Data or elsewhere.

Contract Data should always be amended to allow for a reference to the Employer's Requirements, Contract Data or elsewhere.

Sub-paragraph (a) ignores the existence of elements of the Works divided into parts. This should be revised as follows:

> *"The Commencement Date and the time for Completion of the Works, each Section (if any) and each Part (if any);"*

In sub-paragraph (b) the reference to Contract Data should be amended to include "Employer's Requirements or elsewhere".

In sub-paragraph (d) there is a reference to review periods specified in the Employer's Requirements without mentioning the alternative of Contract Data or elsewhere.

In sub-paragraph (g) there is again a reference to Employer's Requirements which needs to be amended to allow for the alternative of Contract Data or elsewhere.

Paragraph k (iii) needs amending as follows:

> *"Details showing the Contractor's reasonable estimate of the number of each class of the Contractor's Personnel and of each type of Contractor's Equipment currently on the Site and estimated to be subsequently required on the Site, for each major stage of the execution of the Works."*

This amendment gives the Employer a better understanding of workforce and Contractor's Plant changes. This can be especially important at the initial stages when the Contractor might be underdeploying specific piling or excavating Plant.

The Engineer shall Review the initial programme and each revised programme submitted by the Contractor and may give a Notice to the Contractor stating the extent to which it does not comply with the Contract or ceases to reflect actual progress or is otherwise inconsistent with the Contractor's obligations. If the Engineer gives no such Notice:

within 21 days after receiving the initial programme; or
within 14 days after receiving a revised programme,

the Engineer shall be deemed to have given a Notice of No-objection and the initial programme or revised programme (as the case may be) shall be the Programme.

Any change to this definition would be the addition of the phrase "Employer's Requirements, Contract Data, Contract or elsewhere" in place of the word "Contract" in the first sentence.

It is of little point to tie the hands of the Engineer and/or the Employer by using one phrase when a variety would encompass what is required.

The Contractor shall proceed in accordance with the Programme, subject to the Contractor's other obligations under the Contract. The Employer's Personnel shall be entitled to rely on the Programme when planning their activities.

Again, there is a need to widen the scope of the document referred to as the Contract to include:

> *"... The Employer's Requirements, Contract Data where applicable, Contract and/ or any other document which contains the obligations."*

Nothing in any programme, the Programme or any supporting report shall be taken as, or relieve the Contractor from any obligation to give, a Notice under the Contract.

This direction should be simplified as follows:

> *"Nothing in any programme or any supporting report shall be taken as, or relieve the Contractor from any obligation to give, a Notice under the Contract."*

It is difficult to see what the reference to the Programme adds to the instruction.

If, at any time, the Engineer gives a Notice to the Contractor that the Programme fails (to the extent stated) to comply with the Contract or ceases to reflect actual progress or is otherwise inconsistent with the Contractor's obligations, the Contractor shall within 14 days after receiving this Notice submit a revised programme to the Engineer in accordance with this Sub-Clause.

This is a very important power given to the Engineer to monitor the progress within the existing programme and to keep the Contractor on her/his toes.

8.4 Advance Warning

Each Party shall advise the other and the Engineer, and the Engineer shall advise the Parties, in advance of any known or probable future events or circumstances which may:

(a) adversely affect the work of the Contractor's Personnel;
(b) adversely affect the performance of the Works when completed;
(c) increase the Contract Price; and/or
(d) delay the execution of the Works or a Section (if any).

The keyword in this Sub-Clause is the word "may". This gives discretion to everybody to exercise common sense and to warn all concerned should anything occur or appear likely to occur which would adversely affect the project. The only change would be to the Direction which refers to the Works or Section without mentioning the all-important element being a "part".

Thus (d) would be:

"Delay the execution of the Works, Section (if any) or Part (if any)."

The Engineer may request the Contractor to submit a proposal under Sub-Clause 13.3.2 [*Variation by Request for Proposal*] to avoid or minimise the effects of such event(s) or circumstance(s).

To this sub-paragraph should be added the following:

"In that event the Engineer shall promptly issue a Notice to the Contractor seeking either an estimate and/or a proposal as described above."

8.5 Extension of Time for Completion

The Contractor shall be entitled subject to Sub-Clause 20.2 [*Claims For Payment and/or EOT*] to Extension of Time if and to the extent that Completion for the purposes of Sub-Clause

10.1 [*Taking Over the Works and Sections*] is or will be delayed by any of the following causes:

(a) a Variation (except that there shall be no requirement to comply with Sub-Clause 20.2 [*Claims For Payment and/or EOT*]);
(b) a cause of delay giving an entitlement to EOT under a Sub-Clause of these Conditions;
(c) exceptionally adverse climatic conditions, which, for the purpose of these Conditions, shall mean adverse climatic conditions at the Site which are Unforeseeable having regard to climatic data made available by the Employer under Sub-Clause 2.5 [*Site Data and Items of Reference*] and/or climatic data published in the Country for the geographical location of the Site;
(d) Unforeseeable shortages in the availability of Personnel or Goods (or Employer-Supplied Materials, if any) caused by epidemic or governmental actions; or
(e) any delay, impediment or prevention caused by or attributable to the Employer, the Employer's Personnel, or the Employer's other contractors on the Site.

The introductory sentence to this Sub-Clause needs to be amended to take account of Sub-Clause 10.2 [*Taking over of Parts*].

> *"The Contractor shall be entitled subject to Sub-Clause 20.2 [Claims for Payment and/or EOT] to an extension of time if and to the extent where completion for the purposes of Sub-Clause 10.1 [Taking over the Works and Sections] or Sub-Clause 10.2 [Taking over Parts] is or will be Delayed by any of the following causes: . . ."*

When determining each EOT under Sub-Clause 20.2 [*Claims For Payment and/or EOT*], the Engineer shall review previous determinations under Sub-Clause 3.7 [*Agreement or Determination*] and may increase, but shall not decrease, the total EOT.

In this Sub-Clause the direction in this sub-paragraph to the Engineer to review is strange. Review implies consider, approve or reject with reasons. The process here is for the Engineer to take account of the previous Determinations and thereafter may increase the overall extension of time. The word "review" should be removed and it is substituted by the phrase "take account of". In any event any Engineer will obviously take note of the previous extensions of time and aggregate them.

> If a delay caused by a matter, which is the Employer's responsibility is concurrent with a delay caused by a matter which is the Contractor's responsibility, the Contractor's entitlement to EOT shall be assessed in accordance with the rules and procedures stated in the Special Provisions (if not stated, as appropriate taking due regard of all relevant circumstances).

The reference to all relevant circumstances is rather elastic and is not to be encouraged as the engineer should whenever possible adhere to known and trusted procedures.

It seems very odd to have a reference for assessing concurrent opposing Claims, which may or may not result in an EOT for the Contractor to be governed by rules not contained in these Conditions of Contract. The reference to the Particular Conditions is inappropriate. A clear procedure should be set out for dealing with this exact circumstance.

8.6 Delays Caused by Authorities

> If:
>
> (a) the Contractor has diligently followed the procedures laid down by the relevant legally constituted public authorities or private utility entities in the Country;
> (b) these authorities or entities delay or disrupt the Contractor's work; and
> (c) the delay or disruption was Unforeseeable, then this delay or disruption will be considered as a cause of delay under sub-paragraph (b) of Sub-Clause 8.5 [*Extension of Time for Completion*].

This Sub-Clause lacks sufficient information to enable the Contractor to be fully conversant with the situations when this Sub-Clause might apply. This Sub-Clause seems to be applicable only to coordination between the private utility companies working with the Contractor to provide the utilities to be temporary or Permanent Works. It is easy to imagine situations where the particular private utility company causes unnecessary Delay or disruption to the progress of the

Works. However, to try and claim that this Delay or disruption was unforeseeable is virtually impossible.

8.7 Rate of Progress

If, at any time:

(a) actual progress is too slow to complete the Works or a Section (if any) within the relevant Time for Completion; and/or
(b) progress has fallen (or will fall) behind the Programme (or the initial programme if it has not yet become the Programme) under Sub-Clause 8.3 [*Programme*], other than as a result of a cause listed in Sub-Clause 8.5 [*Extension of Time for Completion*], then the Engineer may instruct the Contractor to submit, under Sub-Clause 8.3 [*Programme*], a revised programme describing the revised methods which the Contractor proposes to adopt in order to expedite progress and complete the Works or a Section (if any) within the relevant Time for Completion.

This Sub-Clause lacks clarity. The Engineer should play a significant role at the outset. That reflects what would happen.

> *"Therefore, an amended Sub-clause would read:*
> *If, at any time the Engineer considers:*
>
> *(a) actual progress is too slow to complete the Works, a Section (if any) and Part (if any) within the relevant Time for Completion; and/or*
> *(b) progress has fallen (or will for) behind the programme (or any subsequent programme) under Sub-Clause 8.3 [Programme],*
>
> *other than as a result of the course listed in Sub-Clause 8.5[Extension of time for Completion], then the Engineer may instruct the Contractor to submit, under Sub-Clause 8.3 [Programme], a revised programme describing the methods which the Contractor proposes to adopt in order to expedite progress and complete the Works, a Section (if any) or a Part (if any) within the relevant Time for Completion."*

The reference to the initial programme as not being the programme could be confusing. **According to Sub-clause 8.2 a programme becomes the Programme only after a No-objection notice.**

Unless the Engineer gives a Notice to the Contractor stating otherwise, the Contractor shall adopt these revised methods, which may require increases in the working hours and/or in the numbers of Contractor's Personnel and/or the Goods, at the Contractor's risk and cost. If these revised methods cause the Employer to incur additional costs, the Employer shall be entitled subject to Sub-Clause 20.2 [*Claims For Payment and/or EOT*] to payment of these costs by the Contractor, in addition to Delay Damages (if any).

There is a lack of any time structure in this sub-paragraph. An amendment as follows would add clarity:

> *"The methods which the Contractor may propose in order to rectify the situation where there is a Time lag in the Programme may require increases in the working hours and/or in the numbers of Contractor's Personnel and/or in his Goods at his risk and cost.*
>
> *The Engineer shall promptly issue a Notice of No-objection to the proposal by the Contractor or a Notice setting out the reasons why the Contractor should not adopt all or any of the proposed methodology.*
>
> *If this proposal, approved by the Engineer, cause the Employer to incur additional costs, he shall be entitled subject to Sub-Clause 20.2 [Claims for Payment and/or EOT] the payment of these costs by the Contractor, in addition to Delay damages (if any)."*

Sub-Clause 13.3.1 [*Variation by Instruction*] shall apply to revised methods, including acceleration measures, instructed by the Engineer to reduce delays resulting from causes listed under Sub-Clause 8.5 [*Extension of Time for Completion*].

This direction is self-explanatory.

8.8 Delay Damages

If the Contractor fails to comply with Sub-Clause 8.2 [*Time for Completion*], the Employer shall be entitled subject to Sub-Clause 20.2 [*Claims For Payment and/or EOT*] to payment of Delay Damages by the Contractor for this default. Delay Damages shall be the amount stated in the Contract Data, which shall be paid for every day, which shall elapse between the relevant Time for Completion and the relevant Date of Completion of the

Works or Section. The total amount due under this Sub-Clause shall not exceed the maximum amount of Delay Damages (if any) stated in the Contract Data.

The second sentence of this first paragraph needs amending to take account of alternative sources of information.

> *"Delay damages shall be the sum stated in the Contract Documents which shall be paid for every day which shall elapse between the relevant Time for Completion and the date stated in the relevant Taking-Over Certificate. The total amount due under this Sub-Clause shall not exceed the maximum amount (if any) of Delay damages stated in the Contract Documents."*

It should be noted that the bracketed phrase "if any" has been moved to address the maximum amount and not the concept of Delay Damages.

These Delay Damages shall be the only damages due from the Contractor for the Contractor's failure to comply with Sub-Clause 8.2 [*Time for Completion*], other than in the event of termination under Sub-Clause 15.2 [*Termination for Contractor's Default*] before Completion of the Works. These Delay Damages shall not relieve the Contractor from the obligation to complete the Works, or from any other duties, obligations or responsibilities, which the Contractor may have under or in connection with the Contract.

This Sub-Clause shall not limit the Contractor's liability for Delay Damages in any case of fraud, gross negligence, deliberate default or reckless misconduct by the Contractor.

This last sentence should be amended to provide the correct preposition:

> *"This Sub-Clause, which would otherwise take account of a cap or limit of the total amount of Delay damages, if any, shall not apply where the Contractor has caused the Delay by reason of fraud, deliberate default or reckless misconduct by him."*

8.9 Employer's Suspension

The Engineer may at any time instruct the Contractor to suspend progress of part or all of the Works, which instruction shall state the date and cause of the suspension.

These Conditions of Contract acknowledge the existence of an element or part of the Works.

Therefore, this sub-paragraph should be amended as follows:

> *"The Engineer may, at any time, instruct the Contractor to suspend the execution of the entire Works, a Section or a Part thereof, which instruction shall state the date and the cause of the suspension."*

During such suspension, the Contractor shall protect, store and secure such part or all of the Works (as the case may be) against any deterioration, loss or damage.

To the extent that the cause of such suspension is the responsibility of the Contractor, Sub-Clauses 8.10 [*Consequences of Employer's Suspension*], 8.11 [*Payment for Plant and Materials after Employer's Suspension*] and 8.12 [*Prolonged Suspension*] shall not apply.

The first sentence of the second paragraph of this Sub-Clause need slight amendment to reflect the amendment in the first paragraph:

> *"During such suspension, the Contractor shall protect, store, where applicable, and secure such part, section or the entire Works (as the case maybe) against any deterioration, loss or damage."*

8.10 Consequences of Employer's Suspension

If the Contractor suffers delay and/or incurs Cost from complying with an Engineer's instruction under Sub-Clause 8.9 [*Employer's Suspension*] and/or from resuming the work under Sub-Clause 8.13 [*Resumption of Work*], the Contractor shall be entitled subject to Sub-Clause 20.2 [*Claims For Payment and/or EOT*] to EOT and/or payment of such Cost Plus Profit.

The Contractor shall not be entitled to EOT, or to payment of the Cost incurred, in making good:

(a) the consequences of the Contractor's faulty or defective Design, workmanship, Plant or Materials; and/or
(b) any deterioration, loss or damage caused by the Contractor's failure to protect, store or secure in accordance with Sub-Clause 8.9 [*Employer's Suspension*].

This second paragraph is not explicit.

A more complete direction would be as follows:

> *"Where all or part of the Delay and/or cost incurred is as a result of making good:*
> *the consequences of the Contractor's faulty or defective Design, workmanship, Plant or Materials; and/or any deterioration, loss or damage caused by the Contractor's Failure to protect, store or secure in accordance with Sub-Clause 8.9 [Employer's Suspension] then the Contractor shall not be entitled to an EOT or payment of the cost incurred for the proportion identified above."*

This amended paragraph makes clear that where only part of the Claim for EOT and/or cost incurred is the responsibility of the Contractor then the Contractor can still succeed as to the balance of the EOT and/or costs incurred.

8.11 Payment for Plant and Materials after Employer's Suspension

> The Contractor shall be entitled to payment of the value (as at the date of suspension instructed under Sub-Clause 8.9 [*Employer's Suspension*]) of Plant and/or Materials, which have not been delivered to Site, if:
>
> (a) the work on Plant, or delivery of Plant and/or Materials, has been suspended for more than 28 days and
>   (i) the Plant and/or Materials were scheduled, in accordance with the Programme, to have been completed and ready for delivery to the Site during the suspension period; and
>   (ii) the Contractor provides the Engineer with reasonable evidence that the Plant and/or Materials comply with the Contract; and
>
> (b) the Contractor has marked the Plant and/or Materials as the Employer's property in accordance with the Engineer's instructions.

The only comment on this Sub-Clause refers to the last sentence. Here there is a reference to marking of the Plant or Materials as Employer's property in accordance with the Engineer's instructions. There is no reference to a Sub-Clause directing the Engineer to give instructions as to marking of the Plant and/or Materials as the Employer's property. It would, in fact, be quite reasonable for the Contractor to proactively mark such Plant and/or Materials as ready for delivery to the Site as Employer's property whether instructed by the Engineer or not.

Therefore, an amended sub-paragraph (b) would be as follows:

> *"The Contractor has marked the Plant and/or Materials as the Employer's property whether in accordance with the Engineer's instructions or of his own volition."*

8.12 Prolonged Suspension

If the suspension under Sub-Clause 8.9 [*Employer's Suspension*] has continued for more than 84 days, the Contractor may give a Notice to the Engineer requesting permission to proceed.

If the Engineer does not give a Notice under Sub-Clause 8.13 *[Resumption of Work*] within 28 days after receiving the Contractor's Notice under this Sub-Clause, the Contractor may either:

(a) agree to a further suspension, in which case the Parties may agree the EOT and/or Cost Plus Profit (if the Contractor incurs Cost), and/or payment for suspended Plant and/or Materials, arising from the total period of suspension; or (and if the Parties fail to reach agreement under this sub-paragraph (a))

(b) after giving a (second) Notice to the Engineer, treat the suspension as an omission of the affected part of the Works (as if it had been instructed under Sub-Clause 13.3.1 [*Variation by Instruction*]) with immediate effect including release from any further obligation to protect, store and secure under Sub-Clause 8.9 [*Employer's Suspension*]. If the suspension affects the whole of the Works, the Contractor may give a Notice of termination under Sub-Clause 16.2 [*Termination by Contractor*].

This Sub-paragraph (b) is not clear. What is intended is as follows:

> *"Where the Parties have failed to reach agreement under sub-paragraph (a) then the Contractor may give a Notice to the Engineer that the Contractor will treat the suspended Section, part or entire Works as an omission of the part, section or the entire Works (as if it had been instructed under Sub-Clause 13.3.1 [Variation by Instruction]) with immediate effect including his release from any of further obligation to protect, store or secure under Sub-Clause 8.9 [Employer's Suspension]. If that suspension affects the whole of the Works, the Contractor may give a Notice of Termination under Sub-Clause 16.2 [Termination by Contractor]."*

8.13 Resumption of Work

The Contractor shall resume work as soon as practicable after receiving a Notice from the Engineer to proceed with the suspended work.

At the time stated in this Notice (if not stated, immediately after the Contractor receives this Notice), the Contractor and the Engineer shall jointly examine the Works and the Plant and Materials affected by the suspension. The Engineer shall record any deterioration, loss, damage or defect in the Works or Plant or Materials, which has occurred during the suspension and shall provide this record to the Contractor. The Contractor shall promptly make good all such deterioration, loss, damage or defect so that the Works, when completed, shall comply with the Contract.

These directions need to be consistent with each other. In one line Works, Plant and Materials are treated as alternatives and elsewhere they are treated as a cumulative whole.

An amended instruction would be as follows:

> *"At the time stated in this Notice (if not stated immediately after the Contractor received this Notice) the Contractor and the Engineer shall jointly examine the Works and/or Plant and/or Materials affected by the suspension. The Engineer shall inspect and examine the Works and/or Plant and/or Materials and thereafter record any deterioration, loss, damage or defect and shall provide this record to the Contractor. The Contractor shall promptly make good all such deterioration, loss, damage or defect so that the Works, when completed shall comply with the Contract, Employer's Requirements or elsewhere, and be fit for the purpose(s) intended in accordance with Sub-Clause 4.1 [Contractor's General Obligations]."*

A further instruction to the Contractor would complete the process of resumption of work.

> *"The Contractor, upon completion of the making good shall issue a Notice to the Engineer to that effect. The Engineer shall thereafter promptly issue a Notice No-objection or a Notice stating the reasons why Contractor has not complied with his obligations under this Sub-Clause."*

CHAPTER 9

Clause 9 Tests on Completion

9.1 Contractor's Obligations

The Contractor shall carry out the Tests on Completion in accordance with this Clause and Sub-Clause 7.4 [*Testing by the Contractor*], after submitting the documents under Sub-Clause 5.6 [*As-Built Records*] and Sub-Clause 5.7 [*Operation and Maintenance Manuals*].

The Contractor shall submit to the Engineer, not less than 42 days before the date the Contractor intends to commence the Tests on Completion, a detailed test programme showing the intended timing and resources required for these tests.

This second paragraph does not make any allowance for Tests on Completion for Sections or Tests on Completion of Parts. The way in which this second paragraph is written implies that the detailed test programme applies to the Works, when, in fact, there are likely to be a series of Tests on Completion to allow the Employer to take over Sections or Parts of the Works as the case may be.

A more appropriate instruction to the Contractor would be as follows:

> *"The Contractor shall submit to the Engineer, not later than 42 days before the date the Contractor intends to commence any of the required Tests upon Completion, a detailed test programme showing the intended timing and resources required for each and every Test, as the case may be."*

The Engineer may Review the proposed test programme and may give a Notice to the Contractor stating the extent to which it does not comply with the Contract. Within 14 days after receiving this Notice, the Contractor shall revise the test programme to rectify such non-compliance. If the Engineer gives no such Notice within 14 days after receiving the test programme (or revised test programme), the Engineer shall be deemed to have given a Notice of No-objection. The Contractor shall not commence the Tests on Completion until a Notice of No-objection is given (or is deemed to have been given) by the Engineer.

In addition to any date(s) shown in the test programme, the Contractor shall give a Notice to the Engineer, of not less than 21 days, of the date after which the Contractor will be ready to carry out each of the Tests on Completion. The Contractor shall commence the Tests on Completion within 14 days after this date, or on such day or days as the Engineer shall instruct, and shall proceed in accordance with the Contractor's test programme to which the Engineer has given (or is deemed to have given) a Notice of No-objection.

The only change to this third paragraph within Sub-Clause 9.1 is to move the word "Contract" in the second sentence and replace it with "*the Contract Documents*".

Unless otherwise stated in the Employer's Requirements, the Tests on Completion shall be carried out in stages in the following sequence:

(a) pre-commissioning tests (on or off the Site, as appropriate), which shall include the appropriate inspections and ("dry" or "cold") functional tests to demonstrate that each item of the Works or Section can safely undertake the next stage under sub-paragraph (b) below;

(b) commissioning tests, which shall include the operational tests specified in the Employer's Requirements to demonstrate that the Works or Section can be operated safely and as specified in the Employer's Requirements, under all available operating conditions; and

(c) trial operation (to the extent possible under available operating conditions), which shall demonstrate that the Works or Section perform reliably and in accordance with the Contract.

There needs to be an additional instruction given to the Engineer:

> *"In addition to any dates shown in the test programme, the Contractor shall give a Notice to the Engineer, of not less than 21 days, of the date after which the Contractor will be ready to carry out each of the Tests on Completion. The Engineer may issue a Notice to the Contractor identifying such day or days upon which each of the Test on Completion shall commence. If no such Notice is given then the Contractor shall commence the Tests on Completion within 14 days after the expiry of the period of 21 days and shall proceed in accordance with the Contractor's test program to which the Engineer has given (who is deemed to have given) a Notice of No-objection."*

The changes to this paragraph follow the changes outlined above. All references to Employer's Requirements are amended to "*Contract Documents*" and references "to each item of the Works or section" are amended to: "*Each item of the Works, Section or Part as the case may be*".

> *"Contract Documents"*
> *and references "to each item of the Works or section" are amended to:*
> *"Each item of the Works, Section or Part as the case maybe"*

The tests of each stage described in sub-paragraphs (b) and (c) shall not be commenced until the Works or Section had passed the previous stage.

The only change here is to allow for Parts.

Therefore, the amended instruction is as follows:

> *"The tests of each stage described in sub-paragraphs (b) and (c) above and shall not be commenced until the Works, Section or Part has passed the previous stage."*

During trial operation, when the Works or Section (as the case may be) are operating under stable conditions, the Contractor shall give a Notice to the Engineer that they are ready for any other Tests on Completion, including performance tests. Performance tests shall be carried out to demonstrate whether the Works or Section comply with the performance criteria specified in the Employer's Requirements and with the Schedule of Performance Guarantees.

The same simple amendments are required here which acknowledge the fact that Test on Completion for Parts are part of these Conditions of Contract.

> *"During trial operation, when the Works, Section or Part (as the case may be) are operating under stable Conditions, The Contractor shall give Notice to the Engineer that they are ready for any other Tests on Completion, including performance. Performance tests shall be carried out to demonstrate whether that Works, Section or*

Part (as the case may be) comply with the performance criteria specified in that the Employer's Requirements, Contract Data, Contract or elsewhere in the Employer's Documents and with the Schedule of Performance Guarantees (if any)."

It should be recorded that according to these Conditions of Contract the imposition of Performance Guarantees may not be mandatory.

Trial operation, including performance testing, shall not constitute a taking over under Clause 10 [*Employer's Taking Over*].

Any product produced by, and any revenue or other benefit resulting from, trial operation under this Sub-Clause shall be the property of the Employer.

This Sub-Clause is self-explanatory.

As soon as the Works or Section have, in the Contractor's opinion, passed each stage of the Tests on Completion described in sub-paragraphs (a) to (c) above, the Contractor shall submit a certified report of the results of these tests to the Engineer. The Engineer shall Review each such report and may give a Notice to the Contractor stating the extent to which the results of the tests do not comply with the Contract. If the Engineer does not give such a Notice within 14 days after receiving the results of the tests, the Engineer shall be deemed to have given a Notice of No-objection.

In some instances in these Conditions of Contract the Engineer is permitted to issue a Notice of No-objection in addition to the deeming provision. There is no explanation for the variation.

This is one of the instances when reference to a Part should be given.

Therefore, this sub-paragraph should be rewritten as follows:

"As soon as the Works, Section or Part have passed each stage of the Tests on Completion described in sub-paragraphs (a) to (c) above the Contractor shall submit a certified report of the results of these tests to the Engineer. The Engineer shall review each such report and may give a Notice to the Contractor of No-objection or a Notice stating the extent to which the results of the test do not comply with the Contract. If the Engineer does not give either Notice within 14 days after receiving the results of the tests, the Engineer shall be deemed to have given a Notice of No-objection."

It is at this stage of the execution of the Works that the Contractor has to be very well aware of the number and progress of the Notices applicable to Tests on

Completion which are imminent or outstanding at any one time. Failure to keep a track of such Notices could cause considerable problems for the Contractor subsequently.

In considering the results of the Tests on Completion, the Engineer shall make allowances for the effect of any use of (any part of) the Works by the Employer on the performance or other characteristics of the Works.

This is a placatory concept. It would be very difficult for the Engineer to identify and then quantify the effect of the use of any Part or Section of the Works elsewhere by the Employer on the performance or other characteristics of the particular test being carried out. **Moreover, this is an instruction to the Engineer that is of no interest whatsoever to the Contractor.**

9.2 Delayed Tests

If the Contractor has given a Notice under Sub-Clause 9.1 [*Contractor's Obligations*] that the Works or Section (as the case may be) are ready for Tests on Completion, and these tests are unduly delayed by the Employer's Personnel or by a cause for which the Employer is responsible, Sub-Clause 10.3 [*Interference with Tests on Completion*] shall apply.

There is no opportunity here for the Contractor to give a Notice, which will be essential, that there has been undue Delay. Also, the reference to unduly is insufficiently precise. It might be better to state: "where a test has been delayed for more than 7 days for reasons for which the Employer is responsible then...".

Therefore, this paragraph requires adjustment to take account of that opportunity:

> *"If the Contractor has given the Notice under Sub-Clause 9.1 Contractor's obligations that the Works, Section or Part (as the case maybe) are ready for Test on Completion, and these tests are unduly Delayed by the Employer's Personnel or by a cause for which the Employer is responsible then the Contractor shall issue a Notice to the Engineer identifying the Delay or the cause and if the Engineer after review of the Notice is satisfied and issues a Notice of No-objection then Sub-Clause 10.3 [Interference with Tests on Completion] shall apply."*

There is now a clear procedure to be followed, which all concerned can consider. The procedure identified is fairer to the Employer and yet removes some of the uncertainty from the Contractor by allowing him to take a pragmatic view of the Delay and swiftly issue a Notice identifying the problem. Also, issuing a Notice the Contractor formally draws the attention of the Engineer to the Delay and its cause.

If the Tests on Completion are unduly delayed by the Contractor, the Engineer may by giving a Notice to the Contractor require the Contractor to carry out the tests within 21 days after receiving the Notice. The Contractor shall carry out the tests on such day or days within this period of 21 days as the Contractor may fix, for which the Contractor shall give a prior Notice to the Engineer of not less than 7 days.

This provision identifies the situation where the Delay is the responsibility of the Contractor and quite properly provides a procedure for serving a Notice on the Contractor acknowledging the Delay and instructing the Contractor to take action forthwith. There is absolutely no reason why in the previous paragraph of this Sub-Clause the same opportunities should not be given by the provision of Notices to the Contractor as it is available to the Employer in this paragraph.

There should be an opportunity as is provided earlier in these general Conditions of Contract for the Engineer after receiving a Notice setting out the date or dates when the tests will be carried out to respond stating his preferred dates.

If the Contractor fails to carry out the Tests on Completion within this period of 21 days:

(a) after a second Notice is given by the Engineer to the Contractor, the Employer's Personnel may proceed with the tests;
(b) the Contractor may attend and witness these tests;
(c) within 28 days of these tests being completed, the Engineer shall send a copy of the test results to the Contractor; and
(d) if the Employer incurs additional costs as a result of such testing, the Employer shall be entitled subject to Sub-Clause 20.2 [*Claims For Payment and/or EOT*] to payment by the Contractor of the costs reasonably incurred.

This procedure although logical is not clear.

A suggested amendment is as follows:

> *"If the Contractor fails to carry out the Test on Completion within this period of 21 days or on a date as the Engineer has fixed:*
>
> *(a) a second Notice shall be given by the Engineer to the Contractor stating that the Employer's Personnel may proceed to carry out this particular Test on Completion;*
> *(b) the Employer's Personnel shall give a Notice to the Contractor as to the date when the Test on Completion will be carried out.*
> *(c) the Contractor may attend and witness these tests;*
> *(d) within 28 days of this test being completed the Employer's Personnel shall send certified test reports to the Engineer for review;*
> *(e) the Engineer shall send a copy of these test results to the Contractor; and*
> *(f) the additional costs incurred by the Employer as a result of the Employer's Personnel carrying out such tests shall entitle the Employer subject to Sub-Clause 20.2 [Claims for Payment and/or EOT] to payment by the Contractor of the costs reasonably incurred."*

There is now a provision identifying the contents of the second Notice, identifying the date when the Delayed test shall be carried out, ensuring that the test results are reviewed by the Engineer in this like manner as would occur had the Contractor carried out the tests and finally that the Employer, not conditionally, receives the cost of having to direct his Personnel to carry out the tests.

> Whether or not the Contractor attends, these Tests on Completion shall be deemed to have been carried out in the presence of the Contractor and the results of these tests shall be accepted as accurate.

This paragraph needs no further explanation. If the Contractor cannot bother to attend then providing the Notice did identify the date when the delayed test will be carried out the Contractor has no valid excuse for absence.

9.3 Retesting

> If the Works, or a Section, fail to pass the Tests on Completion, Sub-Clause 7.5 [*Defects and Rejection*] shall apply. The Engineer or the Contractor may require these failed tests, and the Tests on Completion on any related work, to be repeated under the same terms and conditions. Such repeated tests shall be treated as Tests on Completion for the purposes of this Clause.

This proper requirement for retesting needs to be amplified.

The Test has taken place. The certified reports have been reviewed. The Engineer has decided that for the reasons identified the Test has not complied with the standards and performance criteria applicable to it.

It is then for the Contractor, except in the unusual case where the Contractor has unduly delayed, to rectify the reasons/causes why there is a failure of compliance and then by a Notice request a retest of that particular Test on Completion.

There is no reason for there to be any doubt that the test should be repeated. Therefore, the structure of this direction should be less conditional.

The new structure is proposed as follows:

> *"Where the Works, a Section or a Part fail to pass the Test on Completion, Sub-Clause 7.5 [Defects and Rejection] shall apply. The Contractor shall, except where the Employer's Personnel have carried out the test, carry out such Works to meet the reasons set out in the Notice identifying the failure for compliance.*
>
> *Where the Test has been carried out by the Employer's Personnel and a Notice of non-compliance has been issued then a Notice shall be issued to the Contractor to rectify the reasons for non-compliance. If the Contractor fails to carry out the works of rectification under Sub-Clause 7.5 [Defects and Rejection] then in that particular circumstance the Employer may employ persons to carry out these works and the Employer shall thereafter be entitled subject to Sub-Clause 20.2 [Claims for Payment and EOT] to payment by the Contractor of the costs reasonably incurred.*
>
> *Upon completion of these works carried out under Sub-Clause 7.5 [Defects and Rejection] the Contractor or Employer (as the case maybe) will send a Notice to the Engineer stating that the Works, Section or Part are ready for a further Test on Completion. These tests carried out or to be repeated under the same terms and Conditions and shall be treated as Tests on Completion for purposes of this Clause."*

9.4 Failure to Pass Tests on Completion

If the Works, or a Section, fail to pass the Tests on Completion repeated under Sub-Clause 9.3 [*Retesting*], the Engineer shall be entitled to:

(a) order further repetition of Tests on Completion under Sub-Clause 9.3 [*Retesting*];

(b) reject the Works if the effect of the failure is to deprive the Employer of substantially the whole benefit of the Works, in which event the Employer shall have the same remedies as are provided in sub-paragraph (d) of Sub-Clause 11.4 [*Failure to Remedy Defects*];

(c) reject the Section if the effect of the failure is that the Section cannot be used for its intended purpose(s) under the Contract, in which event the Employer shall have the same remedy as is provided in sub-paragraph (c) of Sub-Clause 11.4 [*Failure to Remedy Defects*]; or

(d) issue a Taking-Over Certificate, if the Employer so requests.

The implications of sub-paragraph (b) are not linguistically made clear.

There also needs to be acknowledgement of the existence of a Part of the Works.

Therefore, a proposed amended first paragraph of Sub-Clause 9.4 is as follows:

"Where the Works, a Section or a Part fail to pass the Tests on Completion repeated under Sub-Clause 9.3[Retesting], the Engineer shall be entitled to:

(a) order a further repetition of the Tests on Completion under Sub-Clause 9.3 [Retesting];

(b) if the effect of the Failure is to deprive the Employer of substantially the whole benefit of the Works, Section or Part, and the Contractor is unable or unwilling to rectify the reasons for non-compliance after a Notice requiring him to do so has been issued and the Contractor has failed to respond within seven days, reject the Works, Section or Part (as the case maybe), in which event the Employer shall have the same remedies as are provided in sub-paragraph (d) of Sub-Clause 11.4 [Failure to Remedy Defects]; or

(c) Issue a Taking-Over Certificate in accordance with sub-paragraph (I) of Sub-Clause 10.1 [Taking over the Works and Sections] and Sub-Clause 10. 2 [Taking over Parts], if the Employer so requests."

This amended first paragraph of Sub-Clause 9.4 now contains within it a procedure ensuring that the Contractor is given a Notice identifying that rejection of the Works, Section or Part, as the case may be, is imminent. This procedural addition will not affect the result in the event that the Contractor is unable or unwilling to address the matters set out in the Engineer's Notice stating the reasons for non-compliance. What this additional step does is to ensure that all Parties are clear as to the consequences of this failure by the insertion of an additional Notice from the Engineer and a required response or deemed response from the Contractor.

In the event of sub-paragraph (d) above, the Contractor shall then proceed in accordance with all other obligations under the Contract, and the Employer shall be entitled subject to Sub-Clause 20.2 [*Claims For Payment and/or EOT*] to payment by the Contractor or a reduction in the Contract Price as described under sub-paragraphs (b)(i) or (b)(ii) of Sub-Clause 11.4 [*Failure to Remedy Defects*], respectively. This entitlement shall be without prejudice to any other rights the Employer may have, under the Contract or otherwise.

Having reviewed the effect of the last paragraph of this Sub-Clause it appears that the powers given to the Engineer in the previous paragraph are intended to be cumulative rather than in the alternative. First the Engineer can continue to order a repetition of the Tests on Completion. Where there is a record of results which continuously identifies failure of the tests then, subject to the procedures suggested, the Engineer can reject the Works, Section or Part.

If the rejection has taken place then the Employer can request the Engineer to issue a Taking-Over Certificate for the rejected Works, Section or Part with all the consequences that act implies. Therefore, before the Employer is entitled to rely on Sub-Clause 20.2 [*Claims for Payment*] and/or EOT the Employer will have to establish the cost of the Works, Section or Part as a proportion of the Contract price before Sub-Clause 20.2 can be properly implemented.

CHAPTER 10

Clause 10 Employer's Taking Over

10.1 Taking over the Works and Section

Except as stated in Sub-Clause 9.4 [*Failure to Pass Tests on Completion*], Sub-Clause 10.2 [*Taking Over Parts*] and Sub-Clause 10.3 [*Interference with Tests on Completion*], the Works shall be taken over by the Employer when:

(a) the Works have been completed in accordance with the Contract, including the passing of the Tests on Completion and except as allowed in sub-paragraph (i) below;
(b) the Engineer has given (or is deemed to have given) a Notice of No Objection to the As-built Records submitted under sub-paragraph (a) of Sub-Clause 5.6 [*As-Built Records*];
(c) the Engineer has given (or is deemed to have given) a Notice of No-objection to the provisional O&M Manuals for the Works submitted under Sub-Clause 5.7 [*Operation and Maintenance Manuals*];
(d) the Contractor has carried out the training (if any) as described under Sub-Clause 5.5 [*Training*]; and
(e) a Taking-Over Certificate for the Works has been issued, or is deemed to have been issued in accordance with this Sub-Clause.

The only minor amendment that needs to be made is in the sub-paragraph where the reference to "Contract" needs to be amended to "Employer's Requirements, Contract Data, Contract or other Documents".

However, the heading here refers to taking over the Works and Sections.

Other than for a reference in paragraph 2 of this Sub-Clause there is no reference to Sections. Paragraph 2 merely states that if the Works are divided into

Sections the Contractor may apply for a Taking-Over Certificate for each Section. What this Sub-Clause fails to address is that these Conditions of Contract clearly anticipate taking over by the Employer of Sections of the Works. That has been omitted from Sub-Clause 10.1. It is not sufficient to refer indirectly to Taking-Over Certificates in paragraph 2. The first paragraph needs to be redrafted to take account of the ability to take over Sections. The amendment will be as follows:

> *"Except as stated in Sub-Clause 9.4 [Failure to pass Tests on Completion], Sub-Clause 10.2 [Taking over Parts] and Sub-Clause 10.3 [Interference with Tests on Completion], the Works and/or individual Sections, if any, shall be taken over by the Employer when;*
>
> *(a) the Works and/or individual Sections, if any, have been completed in accordance with the Employer's Requirements, Contract, Contract Data or other relevant Documents, including the methods described in Sub-Clause 8.2 [Time for Completion] and except as allowed in sub-paragraph (i) below;*
> *(b) the Engineer has given (or is deemed to have given) a Notice of No-objection to the as-built records submitted under sub-paragraph (a) of Sub-Clause 5.6 [As-built Records];*
> *(c) the Engineer has given (or is deemed to have given) a Notice of No-objection to the provisional O&M manuals for the Works and/or individual Sections (if any) submitted under Sub-Clause 5.7 [Operation and Maintenance Manuals];*
> *(d) the Contractor has carried out the training (if any) as described under Sub-Clause 5.5 [Training];*
> *and*
> *(e) a Taking Over Certificate for the Works and/or individual Sections, if any, has been issued, or is deemed to have been issued in accordance with this Sub-Clause."*

> The Contractor may apply for a Taking-Over Certificate by giving a Notice to the Engineer not more than 14 days before the Works will, in the Contractor's opinion, be complete and ready for taking over. If the Works are divided into Sections, the Contractor may similarly apply for a Taking-Over Certificate for each Section.

This paragraph is uncontentious. However, it was important in guiding the fundamental review of the previous paragraph.

> If any Part of the Works is taken over under Sub-Clause 10.2 [*Taking Over Parts*], the remaining Works or Section shall not be taken over until the conditions described in sub-paragraphs (a) to (e) above have been fulfilled.

It is important to note that neither here nor in Sub-Clause 10.2 [*Taking Over Parts*] is the Contractor given the opportunity to apply for a Taking-Over Certificate for a Part of the Works.

It is recommended that the previous paragraph is expanded as follows:

> *"If the Works are divided into Parts, the Contractor may similarly apply for a Taking-Over Certificate for each Part."*

By adding that option, the Conditions now ensure that the Works, Sections and Parts are all addressed equally in the taking over procedure.

The Engineer shall, within 28 days after receiving the Contractor's Notice, either:

(i) issue the Taking-Over Certificate to the Contractor, stating the date on which the Works or Section were completed in accordance with the Contract, except for any minor outstanding work and defects (as listed in the Taking-Over Certificate) which will not substantially affect the safe use of the Works or Section for their intended purpose (either until or whilst this work is completed and these defects are remedied); or

(ii) reject the application by giving a Notice to the Contractor, with reasons. This Notice shall specify the work required to be done, the defects required to be remedied and/or the documents required to be submitted by the Contractor to enable the Taking-Over Certificate to be issued. The Contractor shall then complete this work, remedy such defects and/or submit such documents before giving a further Notice under this Sub-Clause.

The documents referred to here are simply the provisional O&M Manuals, which are either incomplete or not in a state of being completely informative.

If the Engineer does not issue the Taking-Over Certificate or reject the Contractor's application within this period of 28 days, and if the conditions described in sub-paragraphs (a) to (d) above have been fulfilled, the Works or Section shall be deemed to have been completed in accordance with the Contract on the fourteenth day after the Engineer receives the Contractor's Notice of application and the Taking-Over Certificate shall be deemed to have been issued.

It seems inconsistent to have a deemed Taking-Over Certificate due on the 14th day after the Engineer receives the Contractor's Notice and yet in the previous paragraph where the Taking-Over Certificate is to be issued the Engineer must state the date on which the Works or Section was completed. In order to be consistent, the first sub-paragraph of the previous paragraph should be amended as follows:

> *"issue a Taking-Over Certificate to the Contractor, the date of which shall be 14 days after the application was received being the notional date on which the Works or Section or Part were completed in accordance with the Contract, except for any minor outstanding work or defects (as listed in the Taking-Over Certificate) which will not substantially affect the safe use of the Works or Section or Part for their intended purpose (either until whilst this work is completed and these defects are remedied)."*

The amendment now insures consistency as between the two paragraphs.

10.2 Taking Over Parts

The Engineer may, at the sole discretion of the Employer, issue a Taking-Over Certificate for any Part of the Permanent Works.

This is very important as it is the first time that "part" of the Works is officially recognised. However as has been noted regularly throughout this commentary the fact that a Part may be considered as being regularly ignored.

The Employer shall not use any part of the Works (other than as a temporary measure, which is either specified in the Employer's Requirements or with the prior agreement of the Contractor) unless and until the Engineer has issued a Taking-Over Certificate for this part. However, if the Employer does use any part of the Works before the Taking-Over Certificate is issued the Contractor shall give a Notice to the Engineer identifying such part and describing such use, and:

(a) that Part shall be deemed to have been taken over by the Employer as from the date on which it is used;
(b) the Contractor shall cease to be liable for the care of such Part as from this date, when responsibility shall pass to the Employer; and
(c) the Engineer shall immediately issue a Taking-Over Certificate for this Part, and any outstanding work to be completed (including Tests on Completion) and/or defects to be remedied shall be listed in this certificate.

In these Conditions of Contract, it seems there is no certainty whether to designate Part with a capital letter or not. Now the existence of Part has been acknowledged that it should be granted a capital letter.

After the Engineer has issued a Taking-Over Certificate for a Part, the Contractor shall be given the earliest opportunity to take such steps as may be necessary to carry out the outstanding work (including Tests on Completion) and/or remedial work for any defects listed in the certificate. The Contractor shall carry out these works as soon as practicable and, in any case, before the expiry date of the relevant DNP.

The procedures available to the Engineer when considering an application for taking over of the Works or Section, where applicable, are not available when the taking over applies to a Part. The key element to Sub-Clause 10.2 is demonstrated in the first line where it is made clear that the Engineer may only issue a Taking-Over Certificate for a Part at the sole discretion of the Employer. However, nowhere in the subsequent content of Sub-Clause 10.2 is there a reference to discretion being exercised by the Employer.

The second paragraph directs that the Employer shall, not may, not use any part of the Works other than as a temporary measure or by the Contractor's consent.

The general Conditions are attempting to indicate the use of a discretion in the second sentence of the second paragraph. The use of mandatory terms in the first sentence in view of the discretion identified in the first paragraph should be removed.

An amended second paragraph should be as follows:

"The Employer may not use any part of the Works (other than as a temporary measure, which is either specified in the Employer's Requirements, Contract Data, Contract or any other document or with the prior agreement of the Contractor) unless and until the Engineer has issued a Taking-Over Certificate for this Part. However, if the Employer exercises his discretion to use any part of the work before the Taking-Over Certificate is issued the Contractor shall give Notice to the Engineer identifying such Part and describing such use,

and

(a) that Part shall be deemed to have been taken over by the Employer as from the date on which it was used;

(b) the Contractor shall cease to be liable for the care of such part as from this date, when responsibility shall pass to the Employer; and

(c) the Engineer shall immediately issue a Taking-Over Certificate for this Part, and any outstanding work to be completed (including Tests on Completion) and/or defects to be remedied shall be listed in this Certificate."

If the Contractor incurs Cost as a result of the Employer taking over and/or using a Part, the Contractor shall be entitled subject to Sub-Clause 20.2 [*Claims For Payment and/or EOT*] to payment of such Cost Plus Profit.

If the Engineer issues a Taking-Over Certificate for any part of the Works, or if the Employer is deemed to have taken over a Part under sub-paragraph (a) above, for any period of delay after the date under sub-paragraph (a) above, the Delay Damages for completion of the remainder of the Works shall be reduced. Similarly, the Delay Damages for the remainder of the Section (if any) in which this Part is included shall also be reduced. This reduction shall be calculated as the proportion, which the value of the Part (except the value of any outstanding works and/or defects to be remedied) bears to the value of the Works or Section (as the case may be) as a whole. The Engineer shall proceed under Sub-Clause 3.7 [*Agreement or Determination*] to agree or determine this reduction (and, for the purpose of Sub-Clause 3.7.3 [*Time limits*], the date the Engineer receives the Contractor's Notice under this Sub-Clause shall be the date of commencement of the time limit for agreement under Sub-Clause 3.7.3).

The provisions of this paragraph shall only apply to the daily rate of Delay Damages, and shall not affect the maximum amount of these damages.

This paragraph is not easy to understand. An introduction should be made indicating that the situation has arisen where Delay Damages can be imposed, then the following procedure can take place.

The suggested change is as follows:

> *"Where the Contractor is liable for Delay Damages as a result of a breach of one or more of these Conditions of Contract then the Part deemed to be taken over by the Employer shall no longer be subject to the imposition of those Delay Damages.*
>
> *Where the Employer is deemed to have taken over a part of the Works under sub-paragraph (a) above, the Delay Damages for completion of the remainder of the Works or Section or Part, as the case may be, shall be reduced pro rata.*
>
> *This reduction shall be calculated as the proportion which the value of the Part (excepting the value of any outstanding Works and/or defects to be remedied) as to the value of the Works, Section or Part, (as the case maybe) to the whole.*
>
> *The Engineer shall proceed under Sub-Clause 3.7 [Agreement or Determination] to agree or determine this reduction (and, for the purpose of Sub-Clause 3.7.3 [Time Limits], the date upon which the Engineer receives the Contractor's Notice under this Sub-Clause show be the date of commencement of the first time limit under Sub-Clause 3.7.3).*
>
> *The provisions of this paragraph shall only apply to the daily rate of Delay Damages, and shall not affect the maximum amount of these damages."*

10.3 Interference with Tests on Completion

If the Contractor is prevented, for more than 14 days (either a continuous period, or multiple periods which total more than 14 days), from carrying out the Tests on Completion by the Employer's Personnel or by a cause for which the Employer is responsible (including any performance test that is not possible due to available operating conditions during trial operation):

(a) the Contractor shall give a Notice to the Engineer describing such prevention;
(b) the Employer shall be deemed to have taken over the Works or Section (as the case may be) on the date when the Tests on Completion would otherwise have been completed; and
(c) the Engineer shall immediately issue a Taking-Over Certificate for the Works or Section (as the case may be).

It is difficult to see how the Engineer can implement this direction to give a Taking-Over Certificate immediately upon receipt of such a Notice. There surely has to be an interim step that allows the Engineer to be aware that time is running and that the interference, whatever it may be, is likely to have consequences for the Employer. A proposed revision to this first paragraph would be as follows:

> *"Where a Contractor is prevented by the Employer's Personnel or by a cause for which the Employer is responsible for more than 7 days, whether continuous or cumulative,*
>
> *then the Contractor shall give a Notice to the Engineer describing such prevention. The Engineer shall make such investigations as he thinks fit. Where the Contractor is prevented for a total of more than 14 days, a further seven days from the initial Notice, from carrying out the Test on Completion:*
>
> *(a) the Contractor shall give a Notice to the Engineer reiterating such prevention,*
> *(b) the Employer shall be deemed to have taken over the Works, Section or Part (as the case may be) on the date when the Tests on Completion would otherwise have been completed; and*
> *(c) the Engineer shall immediately, if he is satisfied that the act of prevention is as described in the Notice, issue a Taking-Over Certificate for the Works, Section or Part (as the case may be)."*

This revision allows for two matters. Firstly, it puts the Engineer on Notice prior to the 14-day cut-off period that there is on-going prevention. Secondly it ensures that the Engineer must be satisfied that there is an act of prevention and by providing that the Contractor must give a first Notice at the seven-day period the Engineer has the opportunity to investigate the act of prevention.

> After the Engineer has issued this Taking-Over Certificate, the Contractor shall carry out the Tests on Completion as soon as practicable and, in any case, before the expiry date of the DNP. The Engineer shall give a Notice to the Contractor, of not less than 14 days, of the date after which the Contractor may carry out each of the Tests on Completion. Thereafter, Sub-Clause 9.1 [*Contractor's Obligations*] shall apply.
>
> If the Contractor suffers delay and/or incurs Cost as a result of being prevented from carrying out the Tests on Completion, the Contractor shall be entitled subject to Sub-Clause 20.2 [*Claims For Payment and/or EOT*] to EOT and/or payment of such Cost Plus Profit.

This is a typical instruction to the Contractor with little interpretative concerns save the requirement to carry out the Tests as soon as practicable. As to who decides, it appears that the decision will be that of the Contractor. However, the role of decider is given to the Engineer who could make a premature request. The Contractor will wish for Tests to take place as soon as the Works have reached an appropriate stage to merit such a test in spite of the alleged fact that the Employer prevented the Test. Therefore, the Contractor should have the opportunity to refute the Notice on the basis of prematurity.

10.4 Surfaces Requiring Reinstatement

> Except as otherwise stated in the Taking-Over Certificate, a certificate for a Section or Part of the Works shall not be deemed to certify completion of any ground or other surfaces requiring reinstatement.

This direction is self-explanatory.

CHAPTER 11

Clause 11 Defects After Taking Over

11.1 Completion of Outstanding Work and Remedying Defects

In order that the Works and Contractor's Documents, and each Section and/or Part, shall be in the condition required by the Contract (fair wear and tear excepted) by the expiry date of the relevant Defects Notification Period or as soon as practicable thereafter, the Contractor shall:

(a) complete any work which is outstanding on the relevant Date of Completion, within the time(s) stated in the Taking-Over Certificate or such other reasonable time as is instructed by the Engineer; and
(b) execute all work required to remedy defects or damage, of which a Notice is given to the Contractor by (or on behalf of) the Employer on or before the expiry date of the DNP for the Works or Section or Part (as the case may be).

The only change would be to remove the word "Contract" and replace by

"Contract Documents".

If a defect appears or damage occurs during the relevant DNP, a Notice shall be given to the Contractor accordingly, by (or on behalf of) the Employer. Promptly thereafter:

(i) the Contractor and the Employer's Personnel shall jointly inspect the defect or damage;

(ii) the Contractor shall then prepare and submit a proposal for necessary remedial work; and

(iii) the second, third and fourth paragraphs of Sub-Clause 7.5 [*Defects and Rejection*] shall apply.

There are no comments on this paragraph save to state that this is a very simple procedure for the Parties to follow in the event that the defect or damage occurs during the DNP.

11.2 Cost of Remedying Defects

All work under sub-paragraph (b) of Sub-Clause 11.1 [*Completion of Outstanding Work and Remedying Defects*] shall be executed at the risk and cost of the Contractor, if and to the extent that the work is attributable to:

(a) the Design of the Works, other than a part of the Design for which the Employer is responsible (if any);

(b) Plant, Materials or workmanship not being in accordance with the Contract;

(c) improper operation or maintenance which was attributable to matters for which the Contractor is responsible (under Sub-Clause 5.5 [*Training*], Sub-Clause 5.6 [*As-Built Records*] and/or Sub-Clause 5.7 [*Operation and Maintenance Manuals*] or otherwise); or

(d) failure by the Contractor to comply with any other obligation under the Contract.

These requirements which come into play in the event that there has been a defect or damage are shown to be read in the alternative. It is quite possible that the outstanding work is attributable to more than one cause and for the sake of simplicity this first paragraph of that Sub-Clause 11.2 should be read cumulatively as well as in the alternative.

The other alteration that occurs frequently in these Conditions of Contract is the change from the reference to "Contract" to "Employer's Requirements, Contract Data, Contract or other relevant document".

Sub-paragraph (d) is a catch-all category that allows the Employer to place the burden of any damage or defect, save exceptionally where the fault is that of the Employer, clearly upon the Contractor.

It is difficult to know what the phrase "or otherwise" means in some sub-paragraph (c) within the bracket being "under Sub-Clauses 5.5 to 5.7 or otherwise".

If the Contractor considers that the work is attributable to any other cause, the Contractor shall promptly give a Notice to the Engineer and the Engineer shall proceed under Sub-Clause 3.7 [*Agreement or Determination*] to agree or determine the cause (and, for the purpose of Sub-Clause 3.7.3 [*Time limits*], the date of this Notice shall be the date of commencement of the time limit for agreement under Sub-Clause 3.7.3). If it is agreed or determined that the work is attributable to a cause other than those listed above, Sub-Clause 13.3.1 [*Variation by Instruction*] shall apply as if such work had been instructed by the Engineer.

This is a relatively unlikely scenario. It would be more appropriate in the first instance to give the opportunity to the Contractor to give a Notice to the Engineer and the Employer to the effect that the work is attributable to another cause.

Thus a suitable addition would be:

> *"Where and to the extent the work is attributable to any other cause the Contractor may give a Notice promptly to the Engineer setting out the extent to which the work is attributable to another cause.*
>
> *Within seven days of receipt of such Notice the Engineer, after review, shall issue a Notice of No-objection or, in the event that the Engineer does not issue a Notice within the seven days, then a Notice of No-objection shall be deemed to be issued and thereafter Sub-Clause 13.3.1 [Variation by Instruction] shall apply as if such work had been instructed by the Engineer."*

There is now in place a clear set of procedures to enable the Contractor to alert the Engineer where, in the unlikely event, the outstanding work is not attributable to one of the four scenarios outlined in the first paragraph.

11.3 Extension of Defects Notification Period

The Employer shall be entitled to an extension of the DNP for the Works, or a Section or a Part:

(a) if and to the extent that the Works, Section, Part or a major item of Plant (as the case may be, and after taking over) cannot be used for the intended purpose(s) by reason of a defect or damage which is attributable to any of the matters under sub-paragraphs (a) to (d) of Sub-Clause 11.2 [*Cost of Remedying Defects*]; and

(b) subject to Sub-Clause 20.2 [*Claims For Payment and/or EOT*].

However, a DNP shall not be extended by more than a period of two years after the expiry of the DNP stated in the Contract Data.

It is very difficult to see how the sub-paragraph (b) fits into the scenario without additional words.

What is intended is that the Employer shall be able to claim against the Accepted Contract Amount in these circumstances. Therefore, sub-paragraph (b) needs to be phrased as follows:

> *"... under Sub-Clause 20.2 [Claims for Payment and/or EOT] Claims for Payment."*

If delivery and/or erection of Plant and/or Materials was suspended under Sub-Clause 8.9 [*Employer's Suspension*] (other than where the cause of such suspension is the responsibility of the Contractor) or Sub-Clause 16.1 [*Suspension by Contractor*], the Contractor's obligations under this Clause shall not apply to any defects or damage occurring more than two years after the DNP for the Works, of which the Plant and/or Materials form part, would otherwise have expired.

The way this paragraph is drafted it appears to refer to Plant and/or Materials, which have yet to be either delivered or erected. That cannot be a correct interpretation. Surely what is intended is that where there has been a period of time where the delivery or erection was suspended then that period of time shall not be added to the two-year maximum extension period for the DNP. There is no clear or logical reason why the period of extension should not be added to the maximum extension period for the DNP.

11.4 Failure to Remedy Defects

If the remedying of any defect or damage under Sub-Clause 11.1 [*Completion of Outstanding Works and Remedying Defects*] is unduly delayed by the Contractor, a date may be fixed by (or on behalf of) the Employer, on or by which the defect or damage is to be remedied. A Notice of this fixed date shall be given to the Contractor by (or on behalf of) the Employer, which Notice shall allow the Contractor reasonable time (taking due regard of all relevant circumstances) to remedy the defect or damage.

There is nothing contentious in this first paragraph save that the Engineer does not appear to play any role in the process of deciding the date when the Works should be carried out.

If the Contractor fails to remedy the defect or damage by the date stated in this Notice and this remedial work was to be executed at the cost of the Contractor under Sub-Clause 11.2 [*Cost of Remedying Defects*], the Employer may (at the Employer's sole discretion):

(a) carry out the work or have the work carried out by others (including any retesting), in the manner required under the Contract and at the Contractor's cost, but the Contractor shall have no responsibility for this work. The Employer shall be entitled subject to Sub-Clause 20.2 [*Claims For Payment and/or EOT*] to payment by the Contractor of the costs reasonably incurred by the Employer in remedying the defect or damage;
(b) accept the damaged or defective work, in which case the Employer shall be entitled subject to Sub-Clause 20.2 [*Claims For Payment and/or EOT*] to:
  (i) payment of Performance Damages by the Contractor in full satisfaction of this failure; or
  (ii) if there is no Schedule of Performance Guarantees under the Contract, or no applicable Performance Damages, a reduction in the Contract Price. The reduction shall be in full satisfaction of this failure only and shall be in the amount as shall be appropriate to cover the reduced value to the Employer as a result of this failure;

(c) require the Engineer to treat any part of the Works which cannot be used for its intended purpose(s) under the Contract by reason of this failure as an omission, as if such omission had been instructed under Sub-Clause 13.3.1 [*Variation by Instruction*]; or
(d) terminate the Contract as a whole with immediate effect (and Sub-Clause 15.2 [*Termination for Contractor's Default*] shall not apply) if the defect or damage deprives the Employer of substantially the whole benefit of the Works. The Employer shall then be entitled subject to Sub-Clause 20.2 [*Claims For Payment and/or EOT*] to recover from the Contractor all sums paid for the Works, plus financing charges and any costs incurred in dismantling the same, clearing the Site and returning Plant and Materials to the Contractor.

This second paragraph makes clear the various alternatives, which are available to the Employer in the event that the Contractor, after receipt of a Notice, fails to carry out the work of remedying any defects or damage within reasonable time. It is clear that this situation only applies in the case of an undue Delay by the Contractor. The reference to undue Delay is new. The headings should be amended to reflect the fact that Delay is a key element.

There is also no opportunity for the Contractor upon receipt of the Notice to respond by issuing a Notice to the Employer giving reasons why the undue Delay has occurred. Rather, proactively, the Contractor should state why the time allocated by the Engineer is not reasonable. Whether it is necessary to say this explicitly or whether this can be inferred as a simple claim by the Contractor is debatable. Only on the second occasion, if the Engineer/the Employer persists, should this Notice be served.

There could be many reasons for the Delay, some of which might be valid entitling the Contractor to further leniency. Others might be of such a nature as to be merely an excuse. By reason of the punitive nature of some of the options available to the Employer in this second paragraph it does appear necessary to have a procedural opportunity given to the Contractor to object by Notice to the Notice of the Employer and to have that objecting Notice reviewed by the Engineer.

Where the Engineer determines that there is no valid excuse then the options outlined in this paragraph should be immediately available to the Employer. There is a question concerning option (b). The damaged or defective work might not be of such a nature as to incur payment of the Non-performance Damages. Therefore, a revision of this option might be as follows:

> *"Accept the damaged or defective work, in which case the Employer shall be entitled subject to some Clause 20.2 [Claims for Payment and/or EOT] to:*
>
> *(i) where applicable and appropriate in terms of the nature of the damage or defective work payment of Non-performance Damages by the Contractor for this failure; or*
> *(ii) if there is no schedule of Performance Guarantees under the Employer's Requirements, Contract Data, Contract or other relevant document, or where by reason of the nature of the defect or damage there are no applicable Non-performance Damages or where the nature of the damaged or defective work is of such a nature that a reduction in the Contract Price will adequately reflect the value of the failure to remedy the defects or damage then a determination or agreement under Sub-Clause 3.7 will determine an appropriate reduction in the Contract Price. The reduction shall be in full satisfaction of this failure only and shall be in the amount as shall be appropriate to cover the reduced value to the Employer as a result of this failure."*

Option (c) directs that if the Engineer can treat the damage or defect to be of such a nature that part of the Works cannot be used for intended purposes then that unusable part shall be treated as on an omission. Again noting that this Sub-Clause only applies where there has been undue Delay it is still necessary for there to be an opportunity for the Contractor to issue a responding Notice stating why an omission would not be appropriate by reason of the minimal nature of the defect or damage.

The same comments apply to the options set out in Sub-paragraph (d).

This addresses the option to terminate the Contract as a whole with immediate effect. It is necessary for the Contractor to have an opportunity here to issue a responding Notice with reasons to the Engineer and the Employer as to why the defects or damage is not of such a nature as to deprive the Employer of substantially the whole benefit of the Works.

Again recalling that this Sub-Clause only arises where there has been unnecessary Delay it is still draconian in nature. The exercise of discretion by the Employer under sub-paragraph (c) or (d) above shall be without prejudice to any other rights he may have, under the Contract or otherwise.

> The exercise of discretion by the Employer under sub-paragraph (c) or (d) above shall be without prejudice to any other rights the Employer may have, under the Contract or otherwise.

This direction is self-explanatory.

11.5 Remedying of Defective Work Off Site

If, during the DNP, the Contractor considers that any defect or damage in any Plant cannot be remedied expeditiously on the Site the Contractor shall give a Notice, with reasons, to the Employer requesting consent to remove the defective or damaged Plant off the Site for the purposes of repair. This Notice shall clearly identify each item of defective or damaged Plant, and shall give details of:

(a) the defect or damage to be repaired;
(b) the place to which defective or damaged Plant is to be taken for repair;
(c) the transportation to be used (and insurance cover for such transportation);
(d) the proposed inspections and testing off the Site;
(e) the planned duration required before the repaired Plant shall be returned to the Site; and
(f) the planned duration for reinstallation and retesting of the repaired Plant (under Sub-Clause 7.4 [*Testing by the Contractor*] and/or Clause 9 [*Tests on Completion*] if applicable).

It is recommended that the introductory paragraph be amended to provide more details as follows:

> *"If, during the DNP, the Contractor considers that any defects or damage in the Plant cannot be remedied expeditiously on the Site he shall give a Notice, with reasons, to the Employer requesting consent to remove the defective or damaged Plant off the Site for the purposes of repair. This Notice shall clearly identify each defective or damaged component of each item of Plant, and shall give details of ... :"*

The clear identification of each component insures that the Contractor has carried out adequate investigation and also gives reassurance to the Employer as to exactly what is defective or damaged. It is noted that the last sentence of this paragraph states:

> *"The Contractor shall also provide any further details that the Employer may reasonably require."*

It could be argued that this last direction within this paragraph encompasses the proposed change to the introductory paragraph. However, as far as Plant is concerned the clear provision of details at the outset is beneficial.

When the Employer gives consent (which consent shall not relieve the Contractor from any obligation or responsibility under this Clause), the Contractor may remove from the Site such items of Plant as are defective or damaged. As a condition of this consent, the Employer may require the Contractor to increase the amount of the Performance Security by the full replacement cost of the defective or damaged Plant.

This last sentence in this last paragraph requires an additional procedure. The amendment should be as follows to add the requirement for the Employer to direct the Engineer to issue a Notice to this effect:

> *"As a condition of this consent, the Employer shall direct the Engineer to issue a Notice to the Contractor requiring the Contractor to increase the amount of the Performance Security by the full replacement cost of the defective or damaged components of the Plant or the full value of the item of Plant which ever is the less."*

The reference to the Employer requiring the Contractor to do something is vague and should be replaced by a procedure.

11.6 Further Tests after Remedying Defects

Within 7 days of completion of the work of remedying of any defect or damage, the Contractor shall give a Notice to the Engineer describing the remedied Works, Section, Part and/or Plant and the proposed repeated tests (under Clause 9 [*Tests on Completion*] or Clause 12 [*Tests after Completion*], as applicable).

This paragraph makes clear the procedure to be followed was that the defects or damage have been remedied. The continued use of brackets in these Conditions of Contract is not helpful. It is far simpler to remove the brackets and add the word "whether".

Within 7 days after receiving this Contractor's Notice, the Engineer shall give a Notice to the Contractor either:

(a) agreeing with such proposed testing; or
(b) instructing the repeated tests that are necessary to demonstrate that the remedied Works, Section, Part and/or Plant comply with the Contract.

It is quite difficult to determine the reason for the second option. If the intention is to give to the Engineer authority to require additional repeated tests, then that should be stated. It seems obvious that the Contractor will propose in its Notice tests that will demonstrate that the remedied item complies with the Contract. This second option for the Engineer should be revised as follows:

> *"Instructing the* ***Contractor to carry out additional tests that are necessary*** *and identifying those tests in order to demonstrate that the remedied Works, Section, Part, and/or Plant comply with the Employer's Requirements, Contract Data, Contract or other relevant document."*

> If the Contractor fails to give such a Notice within the 7 days, the Engineer may give a Notice to the Contractor, within 14 days after the defect or damage is remedied, instructing the repeated tests that are necessary to demonstrate that the remedied Works, Section, Part and/or Plant comply with the Contract.

The change recommended to option in sub-paragraph (b) above should be made to this paragraph. Thus, the amended paragraph is as follows:

> *"If the Contractor fails to give a Notice described in paragraph 1 within seven days, the Engineer may give a Notice to the Contractor, within 14 days after the defect or damage is remedied, instructing the Contractor that repeated tests are necessary and identifying those tests to demonstrate that the remedied Works, Section, Part, and/or Plant comply with the Employer's Requirements, Contract Data, Contract or other relevant document."*

> All repeated tests under this Sub-Clause shall be carried out in accordance with the terms applicable to the previous tests, except that they shall be carried out at the risk and cost of the Party liable, under Sub-Clause 11.2 [*Cost of Remedying Defects*], for the cost of the remedial work.

This paragraph read as a whole does not make sense. All the repeated tests should be carried out under the same terms and Conditions as the previous tests. The only exception is that these tests shall be carried out at the risk and cost of the Party liable for the cost of the Remedial work, under Sub-Clause 11.2 [*Cost of Remedying Defects*]. Breaking the final paragraph into two sentences makes more sense.

11.7 Right of Access after Taking Over

> Until the date 28 days after issue of the Performance Certificate, the Contractor shall have the right of access to all parts of the

Works and to records of the operation, maintenance and performance of the Works, except as may be inconsistent with the Employer's reasonable security restrictions.

The use of the word "parts" in the first line is not helpful. The word "part" is now an integral element with in this Yellow Book Conditions of Contract. Therefore it is more appropriate to replace the word "part" with "element" in this particular context. However, the direction that the Contractor shall continue to have full access is necessary and appropriate.

Whenever the Contractor intends to access any part of the Works or such records during the relevant DNP:

(a) the Contractor shall request access by giving a Notice to the Employer, describing the parts of the Works and/or records to be accessed, the reasons for such access, and the Contractor's preferred date for access. This Notice shall be given in reasonable time in advance of the preferred date for access, taking due regard of all relevant circumstances including the Employer's security restrictions; and
(b) within 7 days after receiving the Contractor's Notice, the Employer shall give a Notice to the Contractor either:
  (i) stating the Employer's consent to the Contractor's request; or
  (ii) proposing reasonable alternative date(s), with reasons. If the Employer fails to give this Notice within the 7 days, the Employer shall be deemed to have given consent to the Contractor's access on the preferred date stated in the Contractor's Notice.

The same comment as to the use of the word "part" in the paragraph applies. However, the first paragraph insisted that the Contractor would have full access save and except where the access was inconsistent with the Employer's reasonable security restrictions, whereas the second paragraph refers in sub-paragraph (a) to the need for the Contractor to take due regard of all relevant circumstances including the Employer's security restrictions. There is no mention in the first paragraph of a much more all-encompassing scope of circumstances of which the Employer's security restrictions is merely an element.

Therefore, that sub-paragraph (a) of the second paragraph should be amended as follows:

> *"The Contractor shall request access by giving Notice to the Employer, describing the elements of the Works and/or records to be accessed, the reasons for such access, and his preferred date for access. This Notice shall be given in reasonable time in advance of the preferred date for access, taking due regard of the Employer's security restrictions."*

> If the Contractor incurs additional Cost as a result of any unreasonable delay by the Employer in permitting access to the Works or such records by the Contractor, the Contractor shall be entitled subject to Sub-Clause 20.2 [*Claims For Payment and/or EOT*] to payment of any such Cost Plus Profit.

There is no intermediate procedural steps necessary to identify unreasonable Delay. There is a deeming provision in the second sub-paragraph which makes clear if the Employer does not respond to the Contractor this Notice within seven days, consent is deemed to have been given.

The only solution is to identify an element of Employer's interference with the right of access.

Therefore, a proposed intermediate paragraph would be as follows:

> *"Where the Employer fails to give any Notice in response to the Contractor's Notice to the Employer within the seven day period and where the deeming provision shall have come into effect but where the Employer restricts or inhibits the rights of access given to the Contractor either on the preferred date or on such reasonable alternative dates then the Employer shall be treated as unreasonably Delaying the Contractor in providing the right of access given to the Contractor under this Sub-Clause."*

11.8 Contractor to Search

> The Contractor shall, if instructed by the Engineer, search for the cause of any defect, under the direction of the Engineer. The Contractor shall carry out the search on the date(s) stated in the Engineer's instruction or other date(s) agreed with the Engineer.

This is an awkward introductory paragraph without indicating any rationale.

An introductory first sentence might be:

> *"Where the Employer and/or Engineer considers that the performance of an element of the Works does not comply with the Employer's Requirements, Contract Data, Contract or other relevant document but the immediate cause is not clear then the Engineer is empowered to direct the Contractor to search for the cause."*

Unless the defect is to be remedied at the cost of the Contractor under Sub-Clause 11.2 [*Cost of Remedying Defects*], the Contractor shall be entitled subject to Sub-Clause 20.2 [*Claims For Payment and/or EOT*] to payment of the Cost Plus Profit of the search.

There is again an omitted sentence. If there is a search there then should be a discovery, although on occasions there may not be a discoverable cause. Therefore, an addition to the second paragraph should be as follows:

> *"Where the cause of the defect has been identified then unless the defect is to be remedied at the cost of the Contractor under Sub-Clause 11.2 [Cost of remedying Defects], the Contractor shall be entitled subject to Sub-Clause 20.2 [Claims for Payment and/or EOT] the payment of the Costs, plus Profit, of the search."*

If the Contractor fails to carry out the search in accordance with this Sub-Clause, the search may be carried out by the Employer's Personnel. The Contractor shall be given a Notice of the date when such a search will be carried out and the Contractor may attend at the Contractor's own cost. If the defect is to be remedied at the cost of the Contractor under Sub-Clause 11.2 [*Cost of Remedying Defects*], the Employer shall be entitled subject to Sub-Clause 20.2 [*Claims For Payment and/or EOT*] to payment by the Contractor of the costs of the search reasonably incurred by the Employer.

This paragraph is logical and understandable.

11.9 Performance Certificate

Performance of the Contractor's obligations under the Contract shall not be considered to have been completed until the Engineer has issued the Performance Certificate to the Contractor, stating the date on which the Contractor fulfilled the Contractor's obligations under the Contract.

The only change necessary here is to the word "Contract", which should be amended as usual to take account of "Employer's Requirements, Contract Data, Contract and any other relevant document."

The Engineer shall issue the Performance Certificate to the Contractor (with a copy to the Employer and to the DAAB) within 28 days after the latest of the expiry dates of the Defects Notification Periods, or as soon thereafter as the Contractor has:

(a) supplied all the Contractor's Documents, and the Engineer has given (or is deemed to have given) a Notice of No-objection to the as-built records under sub-paragraph (b) of Sub-Clause 5.6 [*As-Built Records*]; and
(b) completed and tested all the Works (including remedying any defects) in accordance with the Contract.

It appears there is an hiatus in sub-paragraph (a) of the second paragraph. Not only does the Contractor have to supply all the Contractor's Documents but in order for the process to be validated the Engineer has to issue a Notice of No-objection, or a deeming provision has to be implemented, under Sub-Clause 5.5.2. That step is omitted above.

Therefore an amendment to sub-paragraph (a) is as follows:

> *" ... supplied all the Contractor's Documents and the Engineer has given (or is deemed to have given) a Notice of No-objection to the Contractor's Documents under Sub-Clause 5.2.2 [Review of Contractor's Documents] and to the As-built Records under sub-paragraph (b) of Sub-Clause 5.6 [As-built records]."*

If the Engineer fails to issue the Performance Certificate within this period of 28 days, the Performance Certificate shall be deemed to have been issued on the date 28 days after the date on which it should have been issued, as required by this Sub-Clause.

Only the Performance Certificate shall be deemed to constitute acceptance of the Works.

This is an unfortunately minimalist direction. The properly expanded direction would be:

> *"The Execution of the Works and/or Plant will only be deemed final and complete when the Engineer has issued the Performance Certificate."*

11.10 Unfulfilled Obligations

After the issue of the Performance Certificate, each Party shall remain liable for the fulfilment of any obligation which remains unperformed at that time. For the purposes of determining the nature and extent of unperformed obligations, the Contract shall be deemed to remain in force.

The last sentence is not helpful. A more explicit statement might be:

> *"For the purposes of this Sub-Clause the Contract shall not be deemed to have expired until all remaining obligations have been performed or until the Engineer had issued a Notice that all outstanding obligations have been performed or had been deemed to have been performed."*

However, in relation to Plant, the Contractor shall not be liable for any defects or damage occurring more than two years after expiry of the DNP for the Plant except if prohibited by law or in any case of fraud, gross negligence, deliberate default or reckless misconduct.

11.11 Clearance of Site

Promptly after the issue of the Performance Certificate, the Contractor shall:

(a) remove any remaining Contractor's Equipment, surplus material, wreckage, rubbish and Temporary Works from the Site;
(b) reinstate all parts of the Site which were affected by the Contractor's activities during the execution of the Works and are not occupied by the Permanent Works; and
(c) leave the Site and the Works in the condition stated in the Employer's Requirements (if not stated, in a clean and safe condition).

Sub-paragraph (b) as written could create concerns for the Contractor. It would be preferable for the Contractor to be able to refer to a set of standards or

Conditions, which would apply to reinstatement. Therefore, a proposed amendment would be:

> *(b) reinstate all parts of the Site which were affected by the Contractor's activities during his execution of the Works and are not occupied by the Permanent Works to the standards and Conditions set out in the Employer's Requirements, Contract Data, Contract and/or other relevant Documents. If no such standards or Conditions are in place then the level of reinstatement shall be that which is reasonable in all the circumstances".*

In addition, the reference to the Employer's Requirements in sub-paragraph (c) should be widened to include the Contract Documents.

> If the Contractor fails to comply with sub-paragraphs (a), (b) and/or (c) above within 28 days after the issue of the Performance Certificate, the Employer may sell (to the extent permitted by applicable Laws) or otherwise dispose of any remaining items and/or may reinstate and clean the Site (as may be necessary) at the Contractor's cost.

It would obviously be wise for the Contractor in most circumstances to carry out the obligations under this Sub-Clause itself.

> The Employer shall be entitled subject to Sub-Clause 20.2 [*Claims For Payment and/or EOT*] to payment by the Contractor of the costs reasonably incurred in connection with, or attributable to, such sale or disposal and reinstating and/or cleaning the Site, less an amount equal to the moneys from the sale (if any).

This is a normal direction.

CHAPTER 12

Clause 12 Tests after Completion

12.1 Procedure for Tests after Completion

Procedure for Tests after Completion

12.1 If Tests after Completion are specified in the Employer's Requirements, this Clause shall apply.

The usual comment here to change Employer's Requirements to:

"Contract Documents."

The Employer shall:

(a) provide all electricity, water, sewage (if applicable), equipment, fuel, consumables, instruments, labour, materials, and suitably qualified, experienced and competent staff, as are necessary to carry out the Tests after Completion efficiently and properly; and

(b) carry out the Tests after Completion in accordance with:

(i) the Employer's Requirements,

(ii) the O&M Manuals to which the Engineer has given (or is deemed to have been given) a Notice of No Objection, under Sub-Clause 5.7 [*Operation and Maintenance Manuals*], and

(iii) such guidance as the Contractor may be required to give during the course of these tests; and in the presence of such Contractor's Personnel as either Party may reasonably request.

There is the usual amendment of Employer's Requirements to be made in sub-paragraph (b)(i) to include a range of options. Also, sub-paragraph (b)(iii) mentions such guidance as the Contractor may be required to give. This first paragraph does not describe the procedure for seeking and obtaining that guidance. There should be inserted here:

> *"The Employer shall, by Notice to the Contractor, identify the scope of the guidance and the date or dates upon which such skilled persons able to provide that guidance should be available."*

There is also missing the direction to the Employer to properly record the readings and provide the results to the Contractor and the Engineer. There should be an addition as follows:

> *"(iv) The Employer shall ensure that the readings and or other results of the Tests after Completion are effectively recorded in a manner capable of independent evaluation."*

> The timing of the Tests after Completion shall be as stated in the Employer's Requirements (if not stated, as soon as is reasonably practicable) after the Works or Section (as the case may be) have been taken over under Clause 10 [*Employer's Taking Over*]. The Engineer shall give a Notice to the Contractor, of not less than 21 days, of the date on which and place at which the Tests after Completion will be carried out. This Notice shall also include a test programme showing the estimated timing for each of such tests. Unless otherwise agreed with the Contractor, these tests shall be carried out on this date.

The first omission is the element of the Works which may defined as parts. Parts are deliberately included in these Conditions of Contract and should be consistently mentioned throughout but are not. There is no requirement in this paragraph for the Engineer to prepare the test programme. Therefore, this paragraph should be amended as follows:

> *"The timing of the Tests after Completion shall be as stated in the Employer's Requirements, Contract Data, Contract and all other relevant Documents (if not stated, as soon as is reasonably practicable) after the Works, Section or Part (as the case may be) have been taken over by the Employer. The Engineer shall give Notice to the Contractor, of not less than 21 days, on the date(s) on which and the place at which the Tests after Completion will be carried out. The Engineer will prepare a test programme. The Notice will include the test programme showing the estimated timing for each of such tests. Unless agreed otherwise with the Contractor, the tests shall be carried out on this date or dates."*

> If the Contractor does not attend at the time and place stated in the Engineer's Notice (or otherwise agreed with the Contractor),

the Employer may proceed with the Tests after Completion, which shall be deemed to have been made in the Contractor's presence, and the Contractor shall be deemed to have accepted the readings as accurate.

This is a very important paragraph with regards to the position of the Contractor and the absolute necessity for the Contractor to ensure that skills and experienced persons are present not only to ensure that the Employer carries out the test effectively but also to represent the Contractor's interests as the tests are carried out.

The results of the Tests after Completion shall be compiled and evaluated by both Parties. Appropriate account shall be taken of the effect of the Employer's prior use of the Works.

There is no requirement in the second paragraph for the Employer and/or its Personnel to make records of the readings taken at the tests, properly report them and subsequently work with the Contractor to evaluate the readings. The last paragraph refers to evaluation. The addition to the second paragraph meets the hiatus in the instructions to the Employer to provide a proper and effective system for recording the readings.

12.2 Delayed Tests

If the Contractor incurs Cost as a result of any unreasonable delay by the Employer in carrying out the Tests after Completion, the Contractor shall be entitled subject to Sub-Clause 20.2 [*Claims For Payment and/or EOT*] to payment of such Cost Plus Profit.

There is missing here a procedure for the Contractor to notify the Employer that the Contractor considers the Delay unreasonable. The preliminary sentence should read as follows:

> *"If the Contractor considers that the Employer is exercising undue Delay in carrying out the Tests after Completion the Contractor shall, by a Notice, notify the Employer with a copy to the Engineer of the dates when the Contractor considers that the Tests should reasonably have been carried out. If the Employer does not respond to the Contractor providing dates when the Tests after Completion should be carried out when the Contractor shall every 14 days issue further Notices identifying*

the periods of Delay until such time as the Employer issues a Notice identifying the dates when the tests will be carried out."

If, for reasons not attributable to the Contractor, a Test after Completion on the Works or any Section cannot be completed during the DNP (or any other period agreed by both Parties), then the Works or Section shall be deemed to have passed this Test after Completion.

The proposed addition gives the Contractor the option to put the Engineer on notice that the Contractor does not consider that it is the cause of the delay.

"Where a Test after Completion on any of the Works, Section or any Part has been Delayed until the expiry of the DNP the Contractor shall issue a Notice with reasons as to why the Delay is not attributable to the Contractor. If, for these reasons not attributable to the Contractor, a Test after Completion on the Works, Section or Part cannot be completed during the DNP (or any other period agreed by both Parties), then the Works, Section or Part shall be deemed to have passed this Test after Completion."

12.3 Retesting

Subject to Sub-Clause 12.4 [*Failure to Pass Tests after Completion*], if the Works, or a Section, fail to pass the Tests after Completion:

(a) sub-paragraph (b) of Sub-Clause 11.1 [*Completion of Outstanding Work and Remedying Defects*] shall apply; and
(b) after remedying any defect or damage, Sub-Clause 11.6 [*Further Tests after Remedying Defects*] shall apply.

This first paragraph establishes the procedures in the event that a test after completion fails. It is necessary to amend this first paragraph as follows:

"Subject to Sub-Clause 12.4 Failure to pass Tests after Completion, if the Works, a Section or a Part fail to pass a Test after Completion:

(a) sub-paragraph (b) of Sub-Clause 11.1 [Completion of Outstanding Work and Remedying Defects] shall apply; and
(b) after remedying any defects or damage, Sub-Clause 11.6 [Further Tests after Remedying Defects] shall apply."

If and to the extent that this failure and retesting are attributable to any of the matters listed in sub-paragraphs (a) to (d) of Sub-Clause 11.2 [*Cost of Remedying Defects*] and cause the Employer to incur additional costs, the Employer shall be entitled subject to Sub-Clause 20.2 [*Claims For Payment and/or EOT*] to payment of these costs by the Contractor.

What is necessary is a linking paragraph between paragraphs 1 and 2.

> *"Where the Employer believes that the failure and retesting are attributable to any of the matters listed in sub-paragraphs (a) to (d) of Sub-Clause 11.2 [Cost of Remedying Defects] the Employer shall, by a Notice, notify the Contractor with the reasons as to which matters the Employer relies upon.*
>
> *If the Contractor does not respond within seven days of the Notice then the Contractor shall be deemed to have issued a Notice of No-objection in response to the Employer's Notice. If the Contractor does object by a Notice within the seven day period the matter shall be referred to the Engineer under Sub-Clause 3.7 [Determination or Agreement]."*

12.4 Failure to Pass Tests after Completion

If:

(a) the Works, or a Section, fail to pass any or all of the Tests after Completion; and
(b) applicable Performance Damages are set out in the Schedule of Performance Guarantees the Employer shall be entitled subject to Sub-Clause 20.2 [*Claims For Payment and/or EOT*] to payment of these Performance Damages by the Contractor in full satisfaction of this failure. If the Contractor pays these Performance Damages to the Employer during the DNP, then the Works or Section shall be deemed to have passed these Tests after Completion.

There is at present no linkage between the failure of the Works, Section or Part to pass any of the Tests after Completion and the implementation of the Non-performance Damages. Therefore, this paragraph should be amended as follows:

> *"If the following Conditions apply, namely:*
>
> *(a) the Works, a Section or Part fail to pass any or all of the Tests after Completion;*
> *(b) that failure triggers the payment of the Non-performance Damages set out in the Schedule of Performance Guarantees (if any); and*
> *(c) the Contractor pays these Non-performance Damages to the Employer during the DNP,*

then the Works, Section or Part shall be deemed to have passed the Test or Tests after Completion."

If the Works, or a Section, fail to pass a Test after Completion and, by giving a Notice to the Employer, the Contractor proposes to make adjustments or modifications to the Works or such Section (including an item of Plant):

(i) the Contractor may be instructed by a Notice given by the Employer that right of access to the Works or Section cannot be given until a time that is convenient to the Employer, which time shall be reasonable;
(ii) the Contractor shall remain liable to carry out the adjustments or modifications and to satisfy this Test, within a reasonable period of receiving the Notice under sub-paragraph (i) above; and
(iii) if the Contractor does not receive a Notice under sub-paragraph (i) above during the relevant DNP, the Contractor shall be relieved of the obligation to make such adjustments or modifications and the Works or Section (as the case may be) shall be deemed to have passed this Test after Completion.

This paragraph does not address the situation where the Contractor has given the Notice mentioned in the first line and the Employer has not responded. A proposed amendment to take account of this scenario is as follows:

"If the Works, a Section or a Part fail to pass a Test after Completion and, by giving a Notice to the Employer, the Contractor proposes to make adjustments or modifications to the Works, Section or Part (including an item of Plant):

(i) the Employer may instruct the Contractor may by Notice as to the date or dates when the adjustments or modifications can be carried out; or
(ii) the Employer may instruct the Contractor by Notice that right of access to the Works, Section or Part cannot be given until the time that is convenient to the Employer, which time shall be reasonable;
(iii) and in that event the Contractor shall remain liable to carry out the adjustments or modifications and to satisfy this test, within a reasonable period of receiving the Notice under sub-paragraph (ii) above
(iv) if the Contractor does not receive a Notice under sub-paragraph (ii) above during the relevant DNP, the Contractor shall be relieved of the obligation to make such adjustments or modifications and the Works, Section or Part (as the case maybe) shall be deemed to have passed the Test after Completion."

If the Contractor incurs additional Cost as a result of any unreasonable delay by the Employer in permitting access to the Works or Section by the Contractor, either to investigate the causes of a failure to pass a Test after Completion or to carry out any adjustments or modifications, the Contractor shall be entitled subject to Sub-Clause 20.2 [*Claims For Payment and/or EOT*] to payment of any such Cost Plus Profit.

This Clause does not give an opportunity to the Contractor to serve Notice applicable to unreasonable Delay. Therefore, the paragraph needs to be inserted to that effect. The proposed additional paragraph shall be follows:

> *"If the Contractor considers that the Employer is unreasonably delaying granting access to the Works or Plant either to investigate the cause of the failure to pass a Test after Completion or to carry out any adjustments or modifications to the Works, Section or Part or Plant which is the subject of the failed test then the Contractor shall serve a Notice on the Employer with reasons identifying the unreasonable Delay or Delays. The Employer then has a period of seven days with in which to respond as to why the Delay is reasonable. The Employer does not respond within the seven day period then it shall be deemed to have issued a Notice of No-objection."*

CHAPTER 13

Clause 13 Variations and Adjustments

13.1 Right to Vary

> Variations may be initiated by the Engineer under Sub-Clause 13.3 [*Variation Procedure*] at any time before the issue of the Taking-Over Certificate for the Works.

That first paragraph is a clear statement of fact and reassures the Contractor that the powers of the Engineer to initiate variations continue until the issuance of the final Taking-Over Certificate for the Works.

> Other than as stated under Sub-Clause 11.4 [*Failure to Remedy Defects*], a Variation shall not comprise the omission of any work, which is to be carried out by the Employer or by others unless otherwise agreed by the Parties.

What this paragraph does not address is the power given to the Employer to omit any work and have that work carried out by others. These Conditions of Contract do not address explicitly when and if the Employer can exercise that power unilaterally other than where the Contractor has terminated the Contract and when and if the Contractor may receive full or partial compensation. This paragraph merely addresses the status of a Variation and what falls within the definition of a Variation. Nor does this paragraph state at what stage the agreement by both Parties shall be reached in order to comply with the contents of the second sentence which addresses the financial entitlements of the Contractor in the specific circumstances of an agreed omission.

> The Contractor shall be bound by each Variation instructed under Sub-Clause 13.3.1 [*Variation by Instruction*], and shall execute the Variation with due expedition and without delay,

unless the Contractor promptly gives a Notice to the Engineer stating (with detailed supporting particulars) that:

(a) the varied work was Unforeseeable having regard to the scope and nature of the Works described in the Employer's Requirements;
(b) the Contractor cannot readily obtain the Goods required for the Variation;
(c) it will adversely affect the Contractor's ability to comply with Sub-Clause 4.8 [*Health and Safety Obligations*] and/or Sub-Clause 4.18 [*Protection of the Environment*];
(d) it will have an adverse impact on the achievement of the Schedule of Performance Guarantees; or
(e) it may adversely affect the Contractor's obligation to complete the Works so that they shall be fit for the purpose(s) for which they are intended under Sub-Clause 4.1 [*Contractor's General Obligations*].

There are various comments that arise from this third paragraph of this Sub-Clause. In the first sentence, the introductory sentence, the Notice to the Engineer must have supporting particulars. Elsewhere in these General Conditions of Contract Notices are frequently required to be reasoned. The only place where particulars might be relevant is in the inability to obtain goods. However, where possible consistency is available throughout these Conditions and this should be amended to:

> *"Unless he promptly gives a reasoned Notice to the Engineer stating ..."*

Sub-paragraph (b) suggests a situation where the Goods cannot readily be obtained. This is a very practical reason for a Contractor to be unable to carry out the instruction. It is presumed that the Contractor will have obtained written commentary from the supplier or suppliers of the goods as to the availability or indeed the cost of the goods. It is suspected that the main focus of this paragraph is on availability. There is no hint as to what "readily" means; possibly within a few days as opposed to months, or from outside the country as opposed to in country. Alternatively, "readily" must refer to the moment of execution and means "at such a moment that execution of the Works shall not be delayed".

Sub-paragraph (c) is subjective and vague; safety of the Works may refer to the execution and to the result. It is the view of the Contractor. If the objection encompasses safety then the onus will be on the Contractor to attempt to demonstrate how safety will be reduced. It will not be an easy position to demonstrate.

Presumably expert opinion will have to be obtained in order to persuade the Engineer that the potential instruction is a risk to safety.

Sub-paragraph (d) indicates "that in the Contractor's opinion were the instruction to be implemented it would have an adverse effect on the progress of the Works". The above phraseology is preferable to a reference to the Schedule of Performance Guarantees, which may not exist. This reason may not be sufficient to influence the Engineer if the overall effect of the instruction is to provide a benefit to the Works.

The final exception addresses the situation where the Contractor considers that the Variation may adversely affect the fitness of purpose of the Works upon completion. This is again quite a subjective opinion unless the Contractor can produce expert evidence that the effect of the variation will be to endanger the completed Works. The Contractor may find that he is not in a position to provide the expert evidence required promptly and may then need to serve a second Notice on the Engineer in a timely manner attaching the expert opinion, if any, to support his initial reliance on sub-paragraphs (a)–(e).

> Promptly after receiving this Notice, the Engineer shall respond by giving a Notice to the Contractor cancelling, confirming or varying the instruction. Any instruction so confirmed or varied shall be taken as an instruction under Sub-Clause 13.3.1 [*Variation by instruction*].

This is a sensible provision.

13.2 Value Engineering

> The Contractor may, at any time, submit to the Engineer a written proposal, which (in the Contractor's opinion) will, if adopted:
>
> (a) accelerate completion;
> (b) reduce the cost to the Employer of executing, maintaining or operating the Works;
> (c) improve the efficiency or value to the Employer of the completed Works; or
> (d) otherwise be of benefit to the Employer.

Whenever the Contractor considers putting forward a proposal under this Sub-Clause then it must be weighed in the balance as to whether the Contractor while achieving one or more of the benefits offered to the Employer by the proposal receives adequate compensation. The proposal will be prepared at the Contractor's cost and as will be seen the benefit of any compensation will be shared between the Parties.

The proposal shall be prepared at the cost of the Contractor and shall include the details as stated in sub-paragraphs (a) to (c) of Sub-Clause 13.3.1 [*Variation by Instruction*].

The Engineer shall, as soon as practicable after receiving such proposal, respond by giving a Notice to the Contractor stating his/her consent or otherwise. The Engineer's consent or otherwise shall be at the sole discretion of the Employer. The Contractor shall not delay any work while awaiting a response.

In the second sentence of this paragraph the Engineer is not required to give reasons for his rejection of the proposal. This direction to the Engineer should be amended as follows:

> *"The Engineer shall, as soon as practicable after receiving such proposal, respond by giving a Notice to the Contractor signifying his consent, which may include reservations as to some insubstantial aspects of the proposal, or a request for further particulars prior to giving his consent or his refusal which shall be reasoned."*

This amendment allows scope to the Engineer to give more guidance to the Contractor to the Engineer's reactions to the proposal. The ability to seek further particulars should be inherent in the procedure and not depend upon the granting of consent. It might be that upon receipt of the further particulars that rejection must be made.

The directions to the Contractor that it must not Delay any work while awaiting a response raises concerns. It would be unfortunate for the Contractor to continue with the Works, which would eventually, were the proposal accepted, be wholly or partially redundant. It would be better to allow discretion to the Engineer to stay on the progress of the Works as follows:

> *"The Contractor shall not Delay any work while awaiting a response, unless the Engineer issues a Notice to the Contractor requiring the Contractor to delay work particularly relevant to the proposal."*

If the Engineer gives his/her consent to the proposal, with or without comments, the Engineer shall then instruct a Variation. Thereafter:

(i) the Contractor shall submit any further particulars that the Engineer may reasonably require; and

(ii) then the third paragraph of Sub-Clause 13.3.1 [*Variation by Instruction*] shall apply, which shall include consideration by the Engineer of the sharing (if any) of the benefit, costs and/or delay between the Parties stated in the Particular Conditions.

In the previous paragraph amendments have been made to provide an option to be given to the Engineer to allow for comments. However, this paragraph requires a procedure for the Engineer to issue a further Notice requiring particulars. In fact, the third paragraph of Sub-Clause 13.3.1 [*Variation by Instruction*] does not encompass the situation of a value Engineering proposal.

Therefore, this paragraph needs to be amended as follows:

> *"Where the Engineer gives his consent to the proposal, with or without comments, the Engineer shall then issue a Variation. Thereafter:*
>
> *(i) the Engineer may reasonably issue a further Notice seeking additional particulars from the Contractor and*
> *(ii) the Contractor shall submit those further particulars promptly; and*
> *(iii) then the third paragraph of Sub-Clause 13.3.1 [Variation by Instruction] amended to allow for the concept of Value Engineering shall apply, which shall include consideration by the Engineer of the sharing (if any) of the benefit, costs and/or Delay between the Parties as stated in the Employer's Requirements, Contract Data, Contract or and other relevant Documents."*

The details of the Parties may not be stated in the particular Conditions and therefore to tie the participant to one particular set of Documents is inappropriate. What is not addressed here is the cost, where appropriate, of removal of any work carried out by the Contractor, directly pertinent to the proposal, which was continued with by reason of the direction to proceed with the work in any event while awaiting a response. It would seem inappropriate for the Contractor to bear that cost, however, without specific guidance from these General Conditions it could appear that the costs are borne by the Contractor in any event.

It is proposed that the last sentence is further amended as follows:

> *"...which shall include consideration by the Engineer of the sharing (If any) of the benefit, costs of the proposal and any costs prior to the proposal receiving consent, and/or Delay between the Parties as identified in the Employer's Requirements, Contract Data, Contract and/or other relevant document."*

13.3 Variation Procedure

Subject to Sub-Clause 13.1 [*Right to Vary*], Variations shall be initiated by the Engineer in accordance with either of the following procedures.

13.3.1 Variation by Instruction

The Engineer may instruct a Variation by giving a Notice (describing the required change and stating any requirements for the recording of Costs) to the Contractor in accordance with Sub-Clause 3.5 [*Engineer's Instructions*].

The Contractor shall proceed with execution of the Variation and shall, within 28 days (or other period proposed by the Contractor and agreed by the Engineer) of receiving the Engineer's instruction, submit to the Engineer detailed particulars including:

(a) a description of the varied work performed or to be performed, including details of the resources and methods adopted or to be adopted by the Contractor;
(b) a programme for its execution and the Contractor's proposal for any necessary modifications (if any) to the Programme according to Sub-Clause 8.3 [*Programme*] and to the Time for Completion; and
(c) the Contractor's proposal for adjustment to the Contract Price, with supporting particulars. Whenever the omission of any work forms part (or all) of a Variation, and if:

    - the Contractor has incurred or will incur cost which, if the work had not been omitted, would have been deemed to be covered by a sum forming part of the Accepted Contract Amount; and
    - the omission of the work has resulted or will result in this sum not forming part of the Contract Price this cost may be included in the Contractor's proposal (and, if so, shall be clearly identified). If the Parties have agreed to the omission of any work which is to be carried out by others, the Contractor's proposal may also include the amount of any loss of profit and other losses and damages suffered (or to be suffered) by the Contractor as a result of the omission.

Thereafter, the Contractor shall submit any further particulars that the Engineer may reasonably require.

There are concerns as to various aspects of this important paragraph. In the first sentence, or first paragraph, the Engineer is given power to instruct a Variation.

However, no option is given to the Contractor to seek further information of the required change demanded by the Variation. Therefore, a proposed amendment is as follows:

> *"The Engineer may instruct a Variation by giving a Notice, describing the required change and such specifications for the change as the Engineer considers appropriate and stating any Requirements for the reporting of costs, to the Contractor in accordance with Sub-Clause 3.5 [Engineer's Instructions.]*
>
> *Thereafter the Contractor shall carry out the Variation promptly save that should the Contractor require further particulars of the proposed variation then the Contractor shall promptly serve a Notice on the Engineer describing the further particulars upon which information is sought. The Engineer shall promptly respond with the particulars requested and the Contractor shall thereafter promptly execute the Variation"*

There is now a procedure in place, which gives an opportunity to the Contractor to clarify concerns arising from the construction without being in breach of these general Conditions.

The next concern addresses the apparent conflict between the information, which the Contractor shall submit to the Engineer and the fact that the Variation has been executed. The Contractor is required to submit a proposed design and/or work to be performed; a programme and any amendments and modifications and, finally, the cost implications. All these crucial aspects will be given to the Engineer after, not before, the execution of the Variation.

It is proposed that the first sentence of the second paragraph is amended as follows:

> *"Before proceeding with the execution of the Variation, or after the execution of the Variation in such circumstances as the Engineer may direct, and, in any case, within 28 days (or such other period proposed by the Contractor and agreed by the Engineer) of receiving the Engineer's instructions, the Contractor shall submit to the Engineer detailed particulars including ..."*

This proposed amendment gives the flexibility required to all Parties concerned. It could well be that the cost implications of the Variation are of such a nature that an Employer properly advised would not wish to implement the Variation. However, if the Variation has already been executed that option has been removed.

Sub-paragraph (c) is very important as far as the Contractor is concerned. The sub-paragraph requires the Contractor to supply the financial implications of the Variation. The main body of the sub-paragraph addresses the situation where the Variation requires in whole or in part and omission of work. In that case where that situation arises and if, in implementing the omission, the Contractor incurs costs that would have otherwise, had the work not being omitted, formed part of the Accepted Contract Amount and where, obviously, the omission results in the sum previously forming part of the Contract Price being reduced these wasted costs must be included in the particulars provided in the Contractor's proposal.

> The Engineer shall then proceed under Sub-Clause 3.7 [*Agreement or Determination*] to agree or determine:
>
> (i) EOT, if any; and/or
> (ii) the adjustments to the Contract Price and the Schedule of Payments, if any (and, for the purpose of Sub-Clause 3.7.3 [*Time limits*], the date the Engineer receives the Contractor's submission under this Sub-Clause (including any requested further particulars) shall be the date of commencement of the time limit for agreement under Sub-Clause 3.7.3). The Contractor shall be entitled to such EOT and/or adjustments to the Contract Price, without any requirement to comply with Sub-Clause 20.2 [*Claims For Payment and/or EOT*].

"The word 'Contract' should be exchanged for Contract Documents."

It is also important to note that the document containing the rates and prices does not need to be headed or entitled "Schedule of Rates and Prices".

> If no Schedule of Rates and Prices is included in the Contract, the adjustments under sub-paragraph (ii) above shall be derived from the Cost Plus Profit of executing the work.
>
> If a Schedule of Rates and Prices is included in the Contract, the following provisions of this Sub-Clause 13.3.1 shall apply to the adjustments under sub-paragraph (ii) above.
>
> For each item of work forming part (or all) of a Variation, the appropriate rate or price for the item shall be the rate or price specified for such item in the Schedule of Rates and Prices or, if there is no such item, the rate or price specified for similar work. However, a new rate or price shall be appropriate for an item of work if no rate or price for this item is specified in the Schedule of Rates and Prices and no specified rate or price is appropriate because the item of work is not of similar character,

or is not executed under similar conditions, as any item in the Contract.

Each new rate or price shall be derived from any relevant rates or prices in the Schedule of Rates and Prices, with reasonable adjustments taking account of all relevant circumstances. If no rates or prices are relevant for the derivation of a new rate or price, it shall be derived from the Cost Plus Profit of executing the work.

Until such time as the adjustments under sub-paragraph (ii) above are agreed or determined, the Engineer shall assess a provisional rate or price for the purposes of Interim Payment Certificates.

These paragraphs that set out a procedure for determining a price are virtually standard in any form of Contract. If the work cannot be ascribed to a similar item then the value of the item should be determined from the amount of cost of the item Plus Profit.

Frequently daywork rates will be used to determine, or assist in determining, where the item is priced on the amount of Cost Plus Profit. The Contractor must be wary of the use of an inappropriate daywork item which would produce a disadvantageous price.

Where the Schedule of Rates and Prices is not included in the Contract an allowance should be made for reference to that Schedule whenever it is to be found. The reference to the "Contract" should again be amended to "Employer's Requirements, Contract Data, Contract and/or any other relevant Documents."

13.3.2 Variation by Request for Proposal

The Engineer may request a proposal, before instructing a Variation, by giving a Notice (describing the proposed change) to the Contractor.

There is no need for a bracket here as the contents of the Notice should be explicit and not subject to conditionality.

The Contractor shall respond to this Notice as soon as practicable, by either:

(a) submitting a proposal, which shall include the matters as described in sub-paragraphs (a) to (c) of Sub-Clause 13.3.1 [*Variation by Instruction*]; or

(b) giving reasons why the Contractor cannot comply (if this is the case), by reference to the matters described in sub-paragraphs (a) to (e) of Sub-Clause 13.1 [*Right to Vary*].

These Conditions of Contract use various words for the context of immediacy. It is suggested that one word is used and that that word is "promptly". This authority given to the Engineer is standard whereby the Engineer has an idea and gives the concept to the Contractor to see if it is practical.

If the Contractor submits a proposal, the Engineer shall, as soon as practicable after receiving it, respond by giving a Notice to the Contractor stating his/her consent or otherwise. The Contractor shall not delay any work whilst awaiting a response.

This paragraph should be amended to take into account the cross-fertilisation of ideas as follows:

> *"If the Contractor submits a proposal, the Engineer shall promptly upon receipt respond with his consent, qualified consent or rejection. The Contractor shall not delay any work whilst awaiting a response unless the Engineer issues an instruction to delay such work."*

There must be an allowance for common sense to prevent the Contractor carrying out work particularly relevant to the proposal where that work is going to be redundant should the proposal be successful.

If the Engineer gives consent to the proposal, with or without comments, the Engineer shall then instruct the Variation. Thereafter, the Contractor shall submit any further particulars that the Engineer may reasonably require and the third paragraph of Sub-Clause 13.3.1 [Variation by Instruction] shall apply.

The same comment applies here as applied earlier. The suggested amendment is as follows:

> *"If the Engineer gives his consent to the proposal with or without reservations, the Engineer shall then instruct the Variation. Should the Engineer provide reservations with his consent the Engineer shall then require the Contractor to submit further particulars of the proposal in order to meet the concerns raised in the reservations. Thereafter, the Contractor shall submit those particulars and Sub-Clause 13.3.1 [Variation by Instruction] shall apply."*

If the Engineer does not give consent to the proposal, with or without comments, and if the Contractor has incurred Cost as a result of submitting it, the Contractor shall be entitled subject to Sub-Clause 20.2 [*Claims For Payment and/or EOT*] to payment of such Cost.

This is a reasonably rational direction that the Contractor should be reimbursed for wasted work.

13.4 Provisional Sums

Each Provisional Sum shall only be used, in whole or in part, in accordance with the Engineer's instructions, and the Contract Price shall be adjusted accordingly. The total sum paid to the Contractor shall include only such amounts for the work, supplies or services to which the Provisional Sum relates, as the Engineer shall have instructed.

This is a very awkward first paragraph. A more logical structure might be as follows:

> *"Only the Engineer can direct how and when each provisional sum shall be used. Provisional sums can only be used, in whole or in part in accordance with those instructions. Whenever Provisional Sums are distributed, in whole or in part, that distribution shall be deducted from the Contract price accordingly."*

For each Provisional Sum, the Engineer may instruct:

(a) work to be executed (including Plant, Materials or services to be supplied) by the Contractor, and for which adjustments to the Contract Price and the Schedule of Payments (if any) shall be agreed or determined under Sub-Clause 13.3.1 [*Variation by Instruction*]; and/or
(b) Plant, Materials, works or services to be purchased by the Contractor, from a nominated Subcontractor (as defined in Sub-Clause 4.5 [*Nominated Subcontractors*]) or otherwise, and for which there shall be included in the Contract Price:
  (i) the actual amounts paid (or due to be paid) by the Contractor; and
  (ii) a sum for overhead charges and profit, calculated as a percentage of these actual amounts by applying the

relevant percentage rate (if any) stated in the applicable Schedule. If there is no such rate, the percentage rate stated in the Contract Data shall be applied.

It is not clear why, in sub-paragraph (a) part of the definition of the work to be executed is based in brackets. In sub-paragraph (b) there is no such reticence. For each category the direction is that the Contractor shall be making expenditure. That expenditure shall be determined either from the actual amounts paid or due to be paid plus a sum for overhead charges and profit. The only concern is that the percentage allocated to overhead charges and profit is only identified if it is in an applicable Schedule. If it is not in the Schedule then the Parties must look to a percentage rate stated in the Contract Data. Yet again it is necessary to amend the reference to "Contract Data" to "Contract Documents."

If the Engineer instructs the Contractor under sub-paragraph (a) and/or (b) above, this instruction may include a requirement for the Contractor to submit quotations from the Contractor's suppliers and/or subcontractors for all (or some) of the items of the work to be executed or Plant, Materials, works or services to be purchased. Thereafter, the Engineer may respond by giving a Notice either instructing the Contractor to accept one of these quotations (but such instruction shall not be taken as an instruction under Sub-Clause 4.5 [*Nominated Subcontractors*]) or revoking the instruction. If the Engineer does not so respond within 7 days of receiving the quotations, the Contractor shall be entitled to accept any of these quotations at the Contractor's discretion.

While the principle is clear the wording leaves something to be desired. What is intended is that possible alternative quotations may be obtained for any proposal. Indeed, put simply, it is stressed that the suppliers and/or Subcontractors are not multiples of the same but separate entities. Therefore, a more accurate instruction would be as follows:

> *"If the Engineer instructs the Contractor under sub-paragraph (a) and/or (b) above, this instruction may include a requirement for the Contractor to obtain and then submit a quotation or quotations from separate suppliers and/ or potential Subcontractors for all, or some, of the items of work to be executed or, where applicable, Plant, Materials, Works or services to be purchased. Thereafter, the Engineer may respond by giving a Notice either instructing the Contractor to accept a particular*

quotation or quotations (but any such instruction shall not be taken as an instruction under Sub-Clause 4.5 [Nominated Subcontractors]) or revoking the Instruction. If the Engineer does not respond within seven days of receiving the quotation(s), then all or any of the quotations may be accepted at the Contractor's discretion. Where the Contractor exercises his discretion to accept all or any of the quotations he shall promptly notify the Engineer of the quotations that he has accepted."

Each Statement that includes a Provisional Sum shall also include all applicable invoices, vouchers and accounts or receipts in substantiation of the Provisional Sum.

The use of the word "substantiation" is not helpful. A simple statement of the Requirements may be as follows:

"Each statement provided under Sub-Clause 14.3 [Application for Interim Payment] that includes a Provisional Sum shall also include all applicable certified true copy invoices, vouchers and accounts or receipts applicable to the sums spent under that Provisional Sum."

13.5 Daywork

If a Daywork Schedule is not included in the Contract, this Sub-Clause shall not apply.

The word "Contract" needs to be amended to "Contract Documents".

For work of a minor or incidental nature, the Engineer may instruct that a Variation shall be executed on a daywork basis. The work shall then be valued in accordance with the Daywork Schedule, and the following procedure shall apply.

The use of a daywork Schedule when minor Works are being executed which do not form part of the Works is common practice. However, some Employers see the use of daywork as an abuse of the Contractual relationship and an excuse to overcharge the Employer. Employers will endeavour to extrapolate appropriate rates for any minor or incidental work from the rates, if any, applicable to other parts of the Contractual obligations, however unrealistic such exercises turn out to be.

Before ordering Goods for such work (other than any Goods priced in the Daywork Schedule), the Contractor shall submit one or more quotations from the Contractor's suppliers and/or subcontractors to the Engineer. Thereafter, the Engineer may instruct the Contractor to accept one of these quotations (but such instruction shall not be taken as an instruction under Sub-Clause 4.5 [*Nominated Subcontractors*]). If the Engineer does not so instruct the Contractor within 7 days of receiving the quotations, the Contractor shall be entitled to accept any of these quotations at the Contractor's discretion.

> *"Thereafter, the Engineer may respond by giving a Notice either instructing the Contractor to accept a particular quotation or quotations (but any such instruction shall not be taken as an instruction under Sub-Clause 4.5 [Nominated Subcontractor's]) or revoking the instruction. If the Engineer does not respond within 7 days of receiving the quotation(s), then all or any of the quotations may be accepted at the Contractor's discretion. Where the Contractor exercises his discretion to accept all or any of the quotations he shall promptly notify the Engineer of the quotations that he has accepted."*

Except for any items for which the Daywork Schedule specifies that payment is not due, the Contractor shall deliver each day to the Engineer accurate statements in duplicate (and one electronic copy), which shall include records (as described under Sub-Clause 6.10 [*Contractor's Records*]) of the resources used in executing the previous day's work.

This direction is not helpful to the Engineer. A more all-encompassing direction would be as follows:

> *"The Contractor shall deliver each day to the Engineer accurate and detailed statements whether in paper and or electronic format at the choice of the Engineer which shall include complete records (as described under Sub-Clause 6.10 [Contractor's Records]) of all the resources used in executing the previous day's work, whether the items are identified in the Daywork Schedule or not. Items not identified in the Daywork Schedule shall be marked appropriately."*

The change of emphasis enables the Engineer to know exactly what has been utilised in each item of the daywork.

One copy of each statement shall, if correct and agreed, be signed by the Engineer and promptly returned to the Contractor.

If not correct or agreed, the Engineer shall proceed under Sub-Clause 3.7 [*Agreement or Determination*] to agree or determine the resources (and, for the purpose of Sub-Clause 3.7.3 [*Time limits*], the date the works which are the subject of the Variation under this Sub-Clause are completed by the Contractor shall be the date of commencement of the time limit for agreement under Sub-Clause 3.7.3).

In the next Statement, the Contractor shall then submit priced statements of the agreed or determined resources to the Engineer, together with all applicable invoices, vouchers and accounts or receipts in substantiation of any Goods used in the daywork (other than Goods priced in the Daywork Schedule).

It is very difficult to determine how the section in brackets in this paragraph relates to the main sentence. The paragraph should be separated into its natural elements as follows:

> *"One copy of each statement shall, if correct and agreed, be signed by the Engineer and promptly returned to the Contractor. If not correct or agreed the Engineer shall proceed under Sub-Clause 3.7 [Agreement or Determination] to agree or determine the resources used. For the purposes of Sub-Clause 3.7.3 [Time Limits] the date the Works which are the subject of the variation under this Sub-Clause are completed shall be the date of commencement of the first time limit under Sub-Clause 3.7.3 [Time Limits]. In the next statement under Sub-Clause 14.3 [Application for Interim Payment], the Contractor shall then submit priced statements of the agreed or determined resources to the Engineer, together with all applicable invoices, vouchers and accounts or receipts as evidence in support of the existence of any Goods used in the Daywork (other than Goods already priced in the Daywork Schedule)."*

Unless otherwise stated in the Daywork Schedule, the rates and prices in the Daywork Schedule shall be deemed to include taxes, overheads and profit. For work of a minor or incidental nature, the Engineer may instruct that a Variation shall be executed on a daywork basis. The work shall then be evaluated in accordance with the daywork Schedule and the following procedure shall apply.

This is a practical direction.

13.6 Adjustments for Changes in Laws

Subject to the following provisions of this Sub-Clause, the Contract Price shall be adjusted to take account of any increase or decrease in Cost resulting from a change in:

(a) the Laws of the Country (including the introduction of new Laws and the repeal or modification of existing Laws);
(b) the judicial or official governmental interpretation or implementation of the Laws referred to in sub-paragraph (a) above;
(c) any permit, permission, license or approval obtained by the Employer or the Contractor under sub-paragraph (a) or (b), respectively, of Sub-Clause 1.13 [*Compliance with Laws*]; or
(d) the requirements for any permit, permission, licence and/or approval to be obtained by the Contractor under sub-paragraph (b) of Sub-Clause 1.13 [*Compliance with Laws*], made and/or officially published after the Base Date, which affect the Contractor in the performance of obligations under the Contract. In this Sub-Clause "change in Laws" means any of the changes under sub-paragraphs (a), (b), (c) and/or (d) above.

There is no requirement for the last sentence in this paragraph. The last sentence of the paragraph requires amendment as follows:

> *"… made and/or officially published after the Base Date, which affects the Contractor in the performance of his obligations under the Contract."*

The obligations identified are not general obligations but are those that affect the obligations of the Contractor and therefore have to be defined as such.

If the Contractor suffers delay and/or incurs an increase in Cost as a result of any change in Laws, the Contractor shall be entitled subject to Sub-Clause 20.2 [*Claims For Payment and/or EOT*] to EOT and/or payment of such Cost.

As the paragraph addressing change of Laws includes a variety of matters not pertaining particularly to change of Laws it is necessary to refer to the entire Sub-Clause rather than the phrase.

Therefore, this paragraph requires amendment as follows:

> *"If the Contractor suffers, or will suffer, Delay and/or incurs, or will incur additional cost as a result of any adjustments for changes in Laws under the first paragraph of this Sub-Clause, it shall be entitled subject to Sub-Clause 20.2 [Claims for Payment and/or EOT] to EOT and/or Payment of such Costs."*

> If there is a decrease in Cost as a result of any change in Laws, the Employer shall be entitled subject to Sub-Clause 20.2 [*Claims For Payment and/or EOT*] to a reduction in the Contract Price.

The same amendment made in the previous paragraph is applicable here.

> *"If there is a decrease in cost as a result of any adjustments for changes in laws under the first paragraph of this Sub-Clause, the Employer shall be entitled subject to Sub-Clause 20.2 [Claims for Payment and/or EOT] to a reduction in the Contract Price."*

> If any adjustment to the execution of the Works becomes necessary as a result of any change in Laws:
>
> (i) the Contractor shall promptly give a Notice to the Engineer, or
> (ii) the Engineer shall promptly give a Notice to the Contractor (with detailed supporting particulars).
>
> Thereafter, the Engineer shall either instruct a Variation under Sub-Clause 13.3.1 [*Variation by Instruction*] or request a proposal under Sub-Clause 13.3.2 [*Variation by Request for Proposal*].

This paragraph requires amendment to reflect the above changes.

> *"If any adjustment to the execution of the Works becomes necessary as a result of any adjustments for changes in laws under the first paragraph of this Sub-Clause:*
>
> *(i) the Contractor may promptly give a Notice to the Engineer (with supporting particulars), or*
> *(ii) the Engineer may promptly give a Notice to the Contractor (with supporting particulars).*
>
> *Thereafter, the Engineer shall either instruct a Variation under Sub-Clause 13.3.1 [Variation by Instruction] or request a proposal under Sub-Clause 13.3.2 [Variation by Request for Proposal]."*

13.7 Adjustments for Changes in Cost

If Schedule(s) of Cost Indexation are not included in the Contract, this Sub-Clause shall not apply.

This is the key direction; if the Contract is silent then there will not be any adjustments to allow for changes in cost. This omission could have a critical effect upon the cash flow of the Contractor where a Material such as sand or concrete or indeed steel were yet again subject to sudden increases in price.

The amounts payable to the Contractor shall be adjusted for rises or falls in the cost of labour, Goods and other inputs to the Works, by the addition or deduction of the amounts calculated in accordance with the Schedule(s) of cost indexation.

This direction is a standard direction imposing price flexibility upon the Contractor.

To the extent that full compensation for any rise or fall in Costs is not covered by this Sub-Clause or other Clauses of these Conditions, the Accepted Contract Amount shall be deemed to have included amounts to cover the contingency of other rises and falls in costs.

This is unreasonable. This Sub-Clause provides the methodology for adjustments. This paragraph does not refer to the additional Sub-Clauses which also address adjustment.

The adjustment to be applied to the amount otherwise payable to the Contractor, as certified in Payment Certificates, shall be calculated for each of the currencies in which the Contract Price is payable. No adjustment shall be applied to work valued on the basis of Cost or current prices.

Until such time as each current cost index is available, the Engineer shall use a provisional index for the issue of Interim Payment Certificates. When a *Current Cost Index* is available, the adjustment shall be recalculated accordingly.

There is no obvious reference to the requirement for the production of "Current" cost indices. Is it to be assumed that this requirement is the duty of the Engineer or the Employer as the Employer it is presumed provided the initial index?

If the Contractor fails to complete the Works within the Time for Completion, adjustment of prices thereafter shall be made using either:

(a) each index or price applicable on the date 49 days before the expiry of the Time for Completion of the Works; or
(b) the current index or price whichever is more favourable to the Employer.

CHAPTER 14

Clause 14 Contract Price and Payment

14.1 The Contract Price

Unless otherwise stated in the Particular Conditions:

(a) the Contract Price shall be the lump sum Accepted Contract Amount and be subject to adjustments, additions (including Cost or Cost Plus Profit to which the Contractor is entitled under these Conditions) and/or deductions in accordance with the Contract;
(b) the Contractor shall pay all taxes, duties and fees required to be paid by the Contractor under the Contract, and the Contract Price shall not be adjusted for any of these costs, except as stated in Sub-Clause 13.6 [*Adjustments for Changes in Laws*];
(c) any quantities which may be set out in a Schedule are estimated quantities and are not to be taken as the actual and correct quantities of the Works which the Contractor is required to execute; and
(d) any quantities or price data which may be set out in a Schedule shall be used for the purposes stated in the Schedule and may be inapplicable for other purposes.

There are proposed amendments as follows:

"Unless otherwise stated in the Employer's Requirements, Contract Data, Contract, Particular Conditions or and other relevant Documents:

(a) The Contract Price shall be the lump sum Accepted Contract Amount which shall be subject to such adjustments, additions and/or deductions (including cost or Cost Plus Profit to which the Contractor is entitled under

these Conditions) as agreed or determined in accordance with these General Conditions of Contract, the Employer's Requirements, the Contract Data, the Contract, the Particular Conditions and/or any other relevant document;

(b) The Contractor shall pay all taxes, duties and fees required to be paid by him under the laws of the country and, where applicable, under the Contract, the Contract Price shall not be adjusted for any of these costs except as stated in Sub-Clause 13.6 [Adjustments for Changes in Laws];

(c) any quantities which may be set out in a document forming part of the Contract are estimated quantities and are not to be taken as the actual and correct quantities of the Works which the Contractor is required to execute; and

(d) any quantities or price data which may be set out in a document forming part of the Contract shall be used for the purposes stated in the document only and may be inapplicable for any other purpose. If the document is silent then the information may be referred to generally."

However, if any part of the Works is to be paid according to quantity supplied or work done, the provisions for measurement and valuation shall be stated in the Particular Conditions. The Contract Price shall be valued accordingly, subject to Adjustments in accordance with the Contract. This paragraph does not read effectively with the concept of the Accepted Contract Amount.

An amended paragraph would state as follows:

> *"However, if in addition to the Accepted Contract Amount, any part of the Works is to be paid according to quantity supplied or work done, the provisions for measurement and valuation will be as stated in the Employer's Requirements, Contract Data, Contract, Particular Conditions, and all other relevant Documents. The Contract Price shall be amended accordingly."*

The emphasis on the integrity of the Accepted Contract Amount and its separation from Works carried out based upon quantity or other factors is important. The Employer will be clearly aware of what is included in the Accepted Contract Amount and also aware of the uncertainty that is related to additional Works carried out according to quantity supplied or work carried out. There is no need to refer to adjustments according to the Contract in the last line of this paragraph.

> However, if any part of the Works is to be paid according to quantity supplied or work done, the provisions for measurement and valuation shall be as stated in the Particular Conditions. The Contract Price shall be valued accordingly, subject to adjustments in accordance with the Contract.

This direction appears to refer to Sub-Clause 13.7 [*Adjustments in Changes of Costs*].

14.2 Advance Payment

If no amount of advance payment is stated in the Contract Data, this Sub-Clause shall not apply.

As always, the reference to Contract Data should be amended to:

> *"Contract Documents"*

It would be unfortunate if there were clear references in other Documents to a total Advance Payment and yet it would be impossible to make that Advance Payment because it was not referred to in the Contract Data.

After receiving the Advance Payment Certificate, the Employer shall make an advance payment, as an interest-free loan for mobilisation and design. The amount of the advance payment and the currencies in which it is to be paid shall be as stated in the Contract Data.

The reference to Contract Data in the last line required amendment. Further the last sentence of this paragraph should be amended as follows:

> *"The amount of the Advance Payment and the Currencies in which it is to be paid shall be as stated in the Contract Data, Employer's Requirements, Contract, Particular Conditions, and any other document that is relevant."*

The word "it" was missing in the original version. The situation where the amount of the Advance Payment is able to be identified within the various Documents but there is silence as to the Currencies suggests that the Employer, in the first instance, should propose making the Advance Payment in the current Currency of the country in which the Works are to be located. In the event that the Parties cannot agree to a Currency then that is an ideal situation for a reference to the Engineer under Sub-Clause 3.7 [*Agreement or Determination*].

14.2.1 Advance Payment Guarantee

The Contractor shall obtain (at the Contractor's cost) an Advance Payment Guarantee in amounts and currencies equal to the advance payment, and shall submit it to the Employer with a copy to the Engineer. This guarantee shall be issued by an entity and from within a country (or other jurisdiction) to which the Employer gives consent, and shall be based on the sample form included in the tender documents or on another form agreed by the Employer (but such consent and/or agreement shall not relieve the Contractor from any obligation under this Sub-Clause).

The first line of the Sub-Clause does not address the situation where the currencies are not identified. The next immediate concern is that the Employer is only required to give consent to the issuing country and not to the entity. Finally, it is not clear what are the implications of the last sub-sentence placed in brackets in this Sub-Clause.

Further if the meaning is certain then it needs to be placed within the text and not within brackets. An amended, the first paragraph should be as follows:

> *"The Contractor shall obtain (at his cost) an Advance Payment Guarantee in the Amounts and, where applicable, Currencies equal to the Advance Payment, and shall submit it to the Employer with a copy to the Engineer. This Guarantee shall be issued by an entity and from within a country, or other jurisdiction, to both of which the Employer has given his consent, and the content of the Guarantee shall be based on the sample form included in the Tender Documents or any other Documents, or on another form agreed by the Employer. The aforementioned consents both to the issuing entity and to the form of the guarantee shall not relieve the Contractor from any obligation that he has under this Sub-Clause."*

The Contractor shall ensure that the Advance Payment Guarantee is valid and enforceable until the Advance Payment has been repaid, but its amount may be progressively reduced by the amount repaid by the Contractor as stated in the Payment Certificates.

There needs to be in addition to the above the following:

> *"Where the Contractor has requested that the Advance Payment Guarantee is reduced in amount then the Contractor shall obtain a copy of the Adjusted Payment Guarantee from the issuing entity and give a copy of that Adjusted Guarantee to the Employer and the Engineer."*

This ensures that proper procedure is followed and that the Advance Payment Guarantee is regularly and officially updated in proportion to the Payments made in the Payment Certificates.

If the terms of the Advance Payment Guarantee specify its expiry date, and the Advance Payment has not been repaid by the date 28 days before the expiry date:

(a) the Contractor shall extend the validity of this guarantee until the Advance Payment has been repaid;
(b) the Contractor shall immediately submit evidence of this extension to the Employer, with a copy to the Engineer; and
(c) if the Employer does not receive this evidence 7 days before the expiry date of this guarantee, the Employer shall be entitled to claim under the guarantee the amount of Advance Payment which has not been repaid.

The procedure set out in this paragraph is clear and does not require any suggested amendments.

When submitting the Advance Payment Guarantee, the Contractor shall include an application (in the form of a Statement) for the Advance Payment.

It would be quite unusual for a Contractor not to submit the application for the Advance Payment having gone to the expense of obtaining the Advance Payment Guarantee.

14.2.2 Advance Payment Certificate

The Engineer shall issue an Advance Payment Certificate for the advance payment within 14 days after:

(a) the Employer has received both the Performance Security and the Advance Payment Guarantee, in the form and issued by an entity in accordance with Sub-Clause 4.2.1 [*Contractor's Obligations*] and Sub-Clause 14.2.1 [*Advance Payment Guarantee*] respectively; and

(b) the Engineer has received a copy of the Contractor's application for the Advance Payment under Sub-Clause 14.2.1 [*Advance Payment Guarantee*].

This direction is clear and self-explanatory.

14.2.3 Repayment of Advance Payment

The Advance Payment shall be repaid through percentage deductions in Payment Certificates. Unless other percentages are stated in the Contract Data:

(a) deductions shall commence in the IPC in which the total of all certified interim payments in the same currency as the Advance Payment (excluding the Advance Payment and deductions and release of retention moneys) exceeds ten percent (10%) of the portion of the Accepted Contract Amount payable in that currency less Provisional Sums; and
(b) deductions shall be made at the amortisation rate of one quarter (25%) of the amount of each IPC (excluding the Advance Payment and deductions and release of retention moneys) in the currencies and proportions of the Advance Payment, until such time as the advance payment has been repaid.

This is a notionally complex formula which is believed could be simplified without doing damage to the concept.

The proposed amendment is as follows:

> *"The Advance Payment shall be repaid by means of percentage deductions from the amount certified in the payment certificates. Unless other percentages as stated in the Employer's Requirements, Contract, Contract Data, Particulars Conditions or any other document:*
>
> *(a) the first deduction shall be made in the IPC where the total of all Certified Interim Payments exceeds 10% of the Accepted Contract Amount less Provisional Sums, these Certified Interim Payments shall exclude the Advance Payment, deductions and repayments of Retention; and*

(b) deductions shall be made on the basis of one quarter (25%) of the amount of the first and subsequent IPCs, less the Advance Payment, deductions and repayments of Retention. These deductions shall be made in the Currencies and reducing proportions of the Advance Payment until such time as the Advance Payment has been repaid."

If the Advance Payment has not been repaid before the issue of the Taking-Over Certificate for the Works, or before termination under Clause 15 [*Termination by Employer*], Clause 16 [*Suspension and Termination by Contractor*] or Clause 18 [*Exceptional Events*] (as the case may be), the whole of the balance then outstanding shall immediately become due and payable by the Contractor to the Employer.

This last paragraph of Sub-Clause 14.2.3 ensures that unless certain unusual events have taken place then at the issue of the Taking-Over Certificate any outstanding balance be immediately repayable by the Contractor.

14.3 Application for Interim Payment

The Contractor shall submit a Statement to the Engineer after the end of the period of payment stated in the Contract Data (if not stated, after the end of each month). Each Statement shall:

(a) be in a form acceptable to the Engineer;
(b) be submitted in one paper-original, one electronic copy and additional paper copies (if any) as stated in the Contract Data; and
(c) show in detail the amounts to which the Contractor considers that the Contractor is entitled, with supporting documents which shall include sufficient detail for the Engineer to investigate these amounts together with the relevant report on progress in accordance with Sub-Clause 4.20 [*Progress Reports*].

The first sentence of this Sub-Clause allows for the absence of information in the Contract Data. This flexibility should be inherent throughout these Conditions of Contract. There needs to be a procedure, albeit simple procedure, for submitting a formatted method statement form to the Engineer for consent. The proposed amendment is as follows:

> *"The Contractor shall submit within seven days of receiving the Letter of Acceptance a draft statement to the Engineer for review. If the Engineer does not issue a Notice of No-objection or a Notice of Objection within seven days the Engineer will have been deemed to have issued a Notice of No-objection.*
>
> *Thereafter the Contractor shall submit a statement to the Engineer after the end of the period payment stated in the Contract Data (if not stated, at the end of each month). Each statement shall:*
>
> *(a) be in the form approved by the Engineer;*
> *(b) be submitted in one paper original copy and such electronic copies as stated in the Contract Data or if not stated as required by the Engineer; and*
> *(c) show in detail the amounts to which the Contractor considers himself to be entitled, with supporting Documents which shall include the relevant report on progress in accordance with Sub-Clause 4.20 [Progress Reports]."*

The Statement shall include the following items, as applicable, which shall be expressed in the various currencies in which the Contract Price is payable, in the sequence listed:

(i) the estimated contract value of the Works executed, and the Contractor's Documents produced, up to the end of the period of payment (including Variations but excluding items described in sub-paragraphs (ii) to (x) below);
(ii) any amounts to be added and/or deducted for changes in Laws under Sub-Clause 13.6 [*Adjustments for Changes in Laws*], and for changes in Cost under Sub-Clause 13.7 [*Adjustments for Changes in Cost*];
(iii) any amount to be deducted for retention, calculated by applying the percentage of retention stated in the Contract Data to the total of the amounts under sub-paragraphs (i), (ii) and (vi) of this Sub-Clause, until the amount so retained by the Employer reaches the limit of Retention Money (if any) stated in the Contract Data;
(iv) any amounts to be added and/or deducted for the advance payment and repayments under Sub-Clause 14.2 [*Advance Payment*];
(v) any amounts to be added and/or deducted for Plant and Materials under Sub-Clause 14.5 [*Plant and Materials intended for the Works*];

(vi) any other additions and/or deductions which have become due under the Contract or otherwise, including those under Sub-Clause 3.7 [*Agreement or Determination*];

(vii) any amounts to be added for Provisional Sums under Sub-Clause 13.4 [*Provisional Sums*];

(viii) any amount to be added for release of Retention Money under Sub-Clause 14.9 [*Release of Retention Money*];

(ix) any amount to be deducted for the Contractor's use of utilities provided by the Employer under Sub-Clause 4.19 [*Temporary Utilities*]; and

(x) the deduction of amounts certified in all previous Payment Certificates.

The changes proposed are as follows:

> *"The Statement shall include the following items, as applicable, which shall be expressed in the various Currencies, where applicable, in which the Contract Price is payable, in the sequence listed:*
>
> *(i) The estimated Contract Value of the Works executed, and the number of the Contractor's Documents produced, up to the end of the Period of Payment (including Variations of excluding items described in sub-paragraphs (ii) to (ix) below);*
>
> *(i) Any amount to be deducted for Retention, calculated by applying the percentage of Retention stated in the Contract Data, or elsewhere, to the total of the amounts under sub- paragraphs (i), (ii) and (vi) of this Sub-Clause, until the amount so retained by the Employer reaches the limit, if any, of the Retention Money stated in the Contract Data or elsewhere."*

The changes are adding flexibility in the introductory paragraph as to Currencies, adding an amplification to the Contractor's Documents in the first sub-paragraph, and amending the reference to Contract Data in the third sub-paragraph to allow flexibility and to change the position of the phrase "if any" to the correct position, which is after the reference to limit.

14.4 Schedule of Payments

If the Contract includes a Schedule of Payments specifying the instalments in which the Contract Price will be paid then, unless otherwise stated in this Schedule:

(a) the instalments quoted in the Schedule of Payments shall be treated as the estimated contract values for the purposes of sub-paragraph (i) of Sub-Clause 14.3 [*Application for Interim Payment*];
(b) Sub-Clause 14.5 [*Plant and Materials intended for the Works*] shall not apply; and
(c) if:

  (i) these instalments are not defined by reference to the actual progress achieved in the execution of the Works; and
  (ii) actual progress is found by the Engineer to differ from that on which the Schedule of Payments was based, then the Engineer may proceed under Sub-Clause 3.7 [*Agreement or Determination*] to agree or determine revised instalments (and, for the purpose of Sub-Clause 3.7.3 [*Time limits*], the date when the difference under sub-paragraph (ii) above was found by the Engineer shall be the date of commencement of the time limit for agreement under Sub-Clause 3.7.3). Such revised instalments shall take account of the extent to which progress differs from that on which the Schedule of Payments was based.

Such revised instalments shall take account of the extent to which progress differs from that on which the Schedule of Payments was based. Unlike other directions in these Conditions of Contract this Sub-Clause appears to anticipate that there may well be a Schedule of Payments other than in the Contract. There is no statement here that if there is no mention of a Schedule of Payments in the Contract this Sub-Clause shall not apply.

However, the various Conditions quoted indicate a further degree of flexibility since they might well have been amended in the actual schedule wherever it might be located. The guidance given suggests further flexibility. The first condition makes clear that the instalments are merely to be estimates for the purposes of an Application for Interim Payment.

The final condition makes clear that if the Engineer is of the opinion that the progress does not match the programme, although the programme is not mentioned, then where the instalments are not defined by reference to actual progress then the Engineer may revise the instalments to equate actual progress.

There is no opportunity for the Contractor to question the initial determination by the Engineer to proceed by Sub-Clause 3.7 [*Agreement or Determination*].

It would be preferable to insert a trigger mechanism for the Engineer as follows:

> *" ... Then the Engineer shall by Notice inform the Contractor that actual progress differs from that progress on which the schedule of payments was based, the Contractor shall promptly respond to the Notice providing such explanation or disagreement as the Contractor considers appropriate then the Engineer may proceed under Sub-Clause 3.7 [Agreement or Determination] to agree or determine revised instalments (and for the purposes of Sub-Clause 3.7.3 [Time Limits]), the date when the difference under sub-paragraph (ii) above was found by the Engineer shall be the date of commencement of the first Time Limit under Sub-Clause 3.7.3."*

If the Contract does not include a Schedule of Payments, the Contractor shall submit non-binding estimates of the payments, which the Contractor expects to become due during each period of three months. The first estimate shall be submitted within 42 days after the Commencement Date. Revised estimates shall be submitted at intervals of three months, until the issue of the Taking-Over Certificate for the Works.

This is a sensible provision. However, it might be more effective for the Contractor to project ahead 12 months rather than three months. The programme will encompass the entire execution of the Works there is no reason why a cash flow projection should not mirror the programme.

14.5 Plant and Materials intended for the Works

If no Plant and/or Materials are listed in the Contract Data for payment when shipped and/or payment when delivered, this Sub-Clause shall not apply.

The usual recommendation for amendment of Contract Data to a wider group of Documents is recommended as follows:

> *"If no Plant and/or Materials are listed in the Contract, Contract Data, Employer's Requirements, Particular Conditions and or any other relevant Documents for payment when shipped and/or payment when delivered, this Sub-Clause shall not apply.*
>
> *Where the Employer has listed Plant and/or Materials, which are to be sourced outside the country in which the Permanent Works are to be located the Contractor will obtain Bank Guarantees for the cost of each item of Plant and/or Material identified by the Employer as requiring such Guarantees."*

The first reference to guarantees comes much later when proving value. There needs to be a requirement at the outset for the Contractor to obtain such guarantees where required.

The Contractor shall include, under sub-paragraph (v) of Sub-Clause 14.3 [*Application for Interim Payment*]:

- an amount to be added for Plant and Materials which have been shipped or delivered (as the case may be) to the Site for incorporation in the Permanent Works; and
- an amount to be deducted when the contract value of such Plant and Materials is included as part of the Permanent Works under sub-paragraph (i) of Sub-Clause 14.3 [*Application for Interim Payment*].

This is virtually an application for an Interim Payment on account when the item is either shipped from the supplier or actually delivered to the Site. It does not yet allow for incorporation into the Permanent Works.

The Engineer shall proceed under Sub-Clause 3.7 [*Agreement or Determination*] to agree or determine each amount to be added for Plant and Materials if the following conditions are fulfilled (and, for the purpose of Sub-Clause 3.7.3 [*Time limits*], the date these conditions are fulfilled shall be the date of commencement of the time limit for agreement under Sub-Clause 3.7.3):

(a) the Contractor has:

(i) kept satisfactory records (including the orders, receipts, Costs and use of Plant and Materials) which are available for inspection by the Engineer;
(ii) submitted evidence demonstrating that the Plant and Materials comply with the Contract (which may include test certificates under Sub-Clause 7.4 [*Testing by the Contractor*] and/or compliance verification documentation under Sub-Clause 4.9.2 [*Compliance Verification System*]) to the Engineer; and
(iii) submitted a statement of the Cost of acquiring and shipping or delivering (as the case may be) the Plant and Materials to the Site, supported by satisfactory evidence;

and either:

(b) the relevant Plant and Materials:

(i) are those listed in the Contract Data for payment when shipped;
(ii) have been shipped to the Country, en route to the Site, in accordance with the Contract; and
(iii) are described in a clean shipped bill of lading or other evidence of shipment, which has been submitted to the Engineer together with:

- evidence of payment of freight and insurance;
- any other documents reasonably required by the Engineer; and
- a written undertaking by the Contractor that the Contractor will deliver to the Employer (prior to submitting the next Statement) a bank guarantee in a form and issued by an entity to which the Employer gives consent (but such consent shall not relieve the Contractor from any obligation in the following provisions of this sub-paragraph), in amounts and currencies equal to the amount due under this Sub-Clause. This guarantee shall be in a similar form to the form described in Sub-Clause 14.2.1 [*Advance Payment Guarantee*] and shall be valid until the Plant and Materials are properly stored on Site and protected against loss, damage or deterioration;

The usual amendments to the reference to Contract Data are recommended. Generally, this Sub-Clause addressing Plant and Materials, which are identified in the Employer's Requirements et cetera is reasonably clear. The Engineer shall review the Interim Payment Application and determine whether or not the amount Claimed for each item of Plant and/or Materials should be added to the Interim Payment Certificate.

It is important to recall that at this time the Plant and/or Material has not been placed in the Permanent Works, it is merely either shipped from its place of manufacture or supply or it has actually arrived on Site.

The Schedule of Documents identified, which the Contractor needs to supply to the Engineer are standard Documents. The concern raised earlier as to the wording of the ultimate paragraph which addresses the Bank Guarantee as well as the references in brackets to the consent not relieving the Contractor from any obligation in the following provisions remain.

> *"The Contractor shall obtain (at his cost) an Advance Payment Guarantee in the amounts and, where applicable, Currencies equal to the Advance Payment, and shall submit it to the Employer with a copy to the Engineer. This guarantee shall be issued by an entity and from within a country, or other jurisdiction, to both of which the Employer has given his consent, and the content of the guarantee shall be based on the sample form included in the Tender Documents or any other Documents, or on another form agreed by the Employer. The aforementioned consents both to the issuing entity and to the form of the guarantee shall not relieve the Contractor from any obligation that he has under this Sub-Clause."*

or

(c) the relevant Plant and Materials:

(i) are those listed in the Contract Data for payment when delivered to the Site, and

(ii) have been delivered to and are properly stored on the Site, are protected against loss, damage or deterioration, and appear to be in accordance with the Contract.

The amount so agreed or determined shall take account of the evidence and documents required under this Sub-Clause and of the contract value of the Plant and Materials. If sub-paragraph (b) above applies, the Engineer shall have no obligation to certify any payment under this Sub-Clause until the Employer has received the bank guarantee in accordance with sub-paragraph (b)(iii) above. The sum to be certified by the Engineer in an IPC shall be the equivalent of eighty percent (80%) of this agreed or determined amount. The currencies for this certified sum shall be the same as those in which payment will become due when the contract value is included under sub-paragraph (i) of Sub-Clause 14.3 [*Application for Interim Payment*]. At that time, the Payment Certificate shall include the applicable amount to be deducted, which shall be equivalent to, and in the same currencies and proportions as, this additional amount for the relevant Plant and Materials.

There is a concern here that while the first two sentences refer to agreeing or determining the amount to be added for Plant and Materials then determining that the sum certified shall be 80% of that agreed amount the next sentence refers to the Currencies for this Certified Sum being the same as those in which the payment will become due when the Contract Value is included in the Application for Interim Payment. The Contract Value is the sum agreed or determined before any certification at an 80% level.

In fact the reference to "Contract value" should be replaced by in reference to the "sum certified" so that there is relative cohesion between the procedures set out in this paragraph.

14.6 Issue of IPC

No amount will be certified or paid to the Contractor until:

(a) the Employer has received the Performance Security in the form, and issued by an entity, in accordance with Sub-Clause 4.2.1 [*Contractor's obligations*]; and
(b) the Contractor has appointed the Contractor's Representative in accordance with Sub-Clause 4.3 [*Contractor's Representative*].

This introduction refers to the requirement for the Contractor to receive approval from the Employer not only as to the design and format of the Performance Security but also as to the issuing entity. Further that no funds will be forthcoming including the Advance payment until the Security has been provided. There is a new provision that demands that the Contractor shall appoint the Contractor's Representative before any funds are released.

14.6.1 The IPC

The Engineer shall, within 28 days after receiving a Statement and supporting documents, issue an IPC to the Employer, with a copy to the Contractor:

(a) stating the amount which the Engineer fairly considers to be due; and
(b) including any additions and/or deductions which have become due under Sub-Clause 3.7 [*Agreement or Determination*] or under the Contract or otherwise, with

detailed supporting particulars (which shall identify any difference between a certified amount and the corresponding amount in the Statement and give the reasons for such difference).

This is a clear direction to the Engineer and the direction deals with the concept of the additions and deductions, which have become due as a result of a determination or agreement by the Engineer or as a result of a Contractual obligation or otherwise. And it is the reference to "or otherwise" which is very helpful as it encompasses every other option that is available under these Conditions of Contract, Employer's Requirements, Contract Data, Contract, particular Conditions and any other relevant document. There is no need for the bracketed section to be placed in brackets. The direction stated therein forms part of the necessary requirements. The brackets should be removed.

14.6.2 Withholding (amounts in) an IPC

For some reason which is not clear "amounts in" is placed in brackets; the brackets should be removed.

Before the issue of the Taking-Over Certificate for the Works, the Engineer may withhold an IPC in an amount, which would (after retention and other deductions) be less than the minimum amount of IPC (if any) stated in the Contract Data. In this event, the Engineer shall promptly give a Notice to the Contractor accordingly.

What this first paragraph does not make clear is that the act of withholding may take place at any time and not just before the issuing of the Taking-Over Certificate. Thus, an amendment should be as follows:

> *"At any time during the execution of the Works and before issuing the Taking-Over Certificate for the Works, the Engineer may withhold an IPC where the amount would, after retention and other deductions, be less than the minimum amount (if any) of IPC stated in the Contract Data, particulars Conditions, Employer's Requirements or otherwise."*

Note the introductory words give flexibility to the Engineer and then the reordering of the words prior to reference to retention of the reductions make clear what is intended. Finally, the repositioning of the words in brackets "if any" to qualify minimum amount as opposed to the IPC complete the revisions.

An IPC shall not be withheld for any other reason, although:

(a) if any thing supplied or work done by the Contractor is not in accordance with the Contract, the estimated cost of rectification or replacement may be withheld until rectification or replacement has been completed;
(b) if the Contractor was or is failing to perform any work, service or obligation in accordance with the Contract, the value of this work or obligation may be withheld until the work or obligation has been performed. In this event, the Engineer shall promptly give a Notice to the Contractor describing the failure and with detailed supporting particulars of the value withheld; and/or
(c) if the Engineer finds any significant error or discrepancy in the Statement or supporting documents, the amount of the IPC may take account of the extent to which this error or discrepancy has prevented or prejudiced proper investigation of the amounts in the Statement until such error or discrepancy is corrected in a subsequent Statement.

It is obvious that the first line of this second paragraph needs amendment. The verb "shall" followed by an "although" phrase is a contradiction.

"An IPC may not be withheld for any other reason except:..."

There are incomplete procedures in this paragraph. The second paragraph properly requires the Engineer to give Notice to the Contractor describing the Failure to perform any work service or obligation. However sub-paragraphs (a) and (c) ignore such a requirement.

A Revision is proposed as follows:

(a) If anything supplied or work done by the Contractor is not in accordance with the Contract, Employer's Requirements or otherwise, the estimated cost of rectification or replacement may be withheld until rectification or replacement has been completed. In this event, the Engineer shall promptly give a Notice to the Contractor describing the things supplied or work done which is not in accordance with the Contract or otherwise and identifying with supporting particulars the value withheld; and/or

(b) *If the Contractor was or is failing to perform any work, service or obligation in accordance with the Contract or otherwise, the value of this work or obligation may be withheld until the work or obligation as being performed. In this event, the Engineer shall promptly give a Notice to the Contractor describing the Failure with supporting particulars of the value withheld; and/or*

(c) *If the Engineer finds any significant error or discrepancy in the Statement or supporting Documents, the amount of the IPC may take account of the extent to which this error or discrepancy has prevented or prejudice proper investigation on the amounts in the Statement until error or discrepancy is corrected in a subsequent statement. In this event, the Engineer shall promptly give a Notice to the Contractor describing the error or discrepancy and supporting particulars of the value withheld, if any."*

For each amount so withheld, in the supporting particulars for the IPC the Engineer shall detail his/her calculation of the amount and state the reasons for it being withheld.

This is a repetition of what is stated in the earlier sub-paragraphs in slightly different words and should be deleted.

14.6.3 Correction or modification

The Engineer may in any Payment Certificate make any correction or modification that should properly be made to any previous Payment Certificate. A Payment Certificate shall not be deemed to indicate the Engineer's acceptance, approval, consent or Notice of No-objection to any Contractor's Document or to (any part of) the Work.

This direction does not read easily. A revised version is as follows:

"The Engineer may make any correction or modification in any Payment Certificate to a figure or entry that was erroneous and that should properly have been corrected in a previous Payment Certificate. A Payment Certificate shall not be deemed to indicate the Engineer's acceptance, approval, consent or Notice of No-objection to any Contractor's Document or to any part of the Works."

If the Contractor considers that an IPC does not include any amounts to which the Contractor is entitled, these amounts shall be identified in the next Statement (the "identified amounts" in this paragraph). The Engineer shall then make any correction or modification that should properly be made in the next Payment Certificate.

The location of the bracket is unfortunate.

> *"If the Contractor considers that an IPC does not include any amounts to which he is entitled he shall identify these amounts, the 'identified amounts' described subsequently, in his next Statement. The Engineer shall then make any correction or modification that should be properly made in the next Payment Certificate."*

Thereafter, to the extent that:

(a) the Contractor is not satisfied that this next Payment Certificate includes the identified amounts; and
(b) the identified amounts do not concern a matter for which the Engineer is already carrying out his/her duties under Sub-Clause 3.7 [*Agreement or Determination*]
the Contractor may, by giving a Notice, refer this matter to the Engineer and Sub-Clause 3.7 [*Agreement or Determination*] shall apply (and, for the purpose of Sub-Clause 3.7.3 [*Time limits*], the date the Engineer receives this Notice shall be the date of commencement of the time limit for agreement under Sub-Clause 3.7.3).

This paragraph has a very clear procedure for the Contractor to follow in the event that the Contractor is still concerned as to outstanding identified amounts.

14.7 Payment

The Employer shall pay to the Contractor:

(a) the amount certified in each Advance Payment Certificate within the period stated in the Contract Data (if not stated, 21 days) after the Employer receives the Advance Payment Certificate;
(b) the amount certified in each IPC issued under:
    (i) Sub-Clause 14.6 [*Issue of IPC*], within the period stated in the Contract Data (if not stated, 56 days) after the Engineer receives the Statement and supporting documents; or

(ii) Sub-Clause 14.13 [*Issue of FPC*], within the period stated in the Contract Data (if not stated, 28 days) after the Employer receives the IPC; and

(c) the amount certified in the FPC within the period stated in the Contract Data (if not stated, 56 days) after the Employer receives the FPC.

This key Sub-Clause omits elements in the payment structure. Further paragraph (b) refers to the issue of a final payment certificate. That method cannot be dealt with under the heading referring to each IPC.

A suggested resolution is as follows:

"The Employer shall pay to the Contractor:

(a) the amount certified in each Advanced Payment Certificate, where more than one is applicable, within the period stated in the Contract Data, if not stated, 21 days, after the Employer receives the Advance Payment Certificate;
(b) the amount certified in each IPC issued under:
Sub-Clause 14.6 [Issue of IPC] within the period stated in the Contract Data, if not stated, 56 days, after the Engineer received the Statement and the supporting Documents;
(c) the amount certified in the provisional Final Payment Certificate under Sub-Clause 14.13 [Issue of FPC] within the period stated in the Contract Data, if not stated, 56 days after the Employer receives the PFPC;
(d) the amount of certified under Sub-Clause 14.13 [Issue of FPC] within the period stated in the Contract Data, if not stated, 56 days after the Employer receives the FPC."

Payment of the amount due in each currency shall be made into the bank account, nominated by the Contractor, in the payment country (for this currency) specified in the Contract.

This is not well drafted. It is intended to be a reminder to the Employer that where there are agreed proportions of Currency constituting the Accepted Contract Amount then proportional Currencies shall be paid accordingly. A revision might be as follows:

"Where the Contract Data, Employer's Requirements or/and other Documents require that the Accepted Contract Amount is paid in proportional Currencies then when making each payment the Employer shall ensure that the amount due in each particular Currency is paid into the particular bank account nominated by the Contractor in the payment Country for that particular Currency, whether specified in the Contract or elsewhere."

The first paragraph does not allow for a single Advance Payment thus requiring a single Advance Payment Certificate. Further the sub-paragraphs are cumulative whereas there will be situations where the sub-paragraphs can operate in the alternative especially when the Advance Payment has been paid.

14.8 Delayed Payment

If the Contractor does not receive payment in accordance with Sub-Clause 14.7 [*Payment*], the Contractor shall be entitled to receive financing charges compounded monthly on the amount unpaid during the period of delay. This period shall be deemed to commence on the expiry of the time for payment specified in Sub-Clause 14.7 [*Payment*], irrespective (in the case of sub-paragraph (b) of Sub-Clause 14.7) of the date on which any IPC is issued.

This first paragraph reminds the Contractor that he is entitled to the financing charges incurred by not having received his monthly payment from the Employer.

Unless otherwise stated in the Contract Data, these financing charges shall be calculated at the annual rate of three percent (3%) above:

(a) the average bank short-term lending rate to prime borrowers prevailing for the currency of payment at the place for payment, or
(b) where no such rate exists at that place, the same rate in the country of the currency of payment, or
(c) in the absence of such a rate at either place, the appropriate rate fixed by the law of the country of the currency of payment.

Also obtaining information from the relevant central bank where each different Currency is held or issued as to what is the discounted rate may be somewhat difficult. It would be much fairer to fix the rate at 3% above the rate of interest established by the central bank for day-to-day banking.

The Contractor shall by request be entitled to payment of these financing charges by the Employer without:

(i) the need for the Contractor to submit a Statement or any formal Notice (including any requirement to comply with Sub-Clause 20.2 [*Claims For Payment and/or EOT*]) or certification; and
(ii) prejudice to any other right or remedy.

This is a useful Sub-Clause for the Contractor. The strange aspect is that the Sub-Clause deliberately does not require the Contractor to issue a Notice to either the Employer or Engineer. Also the reference to a "request" appears unprofessional especially as the result of the request, payment of financing charges, could have a significant effect upon both parties. All Contractors should put any form of such "request" into a formal form copied to all concerned.

14.9 Release of Retention Money

After the issue of the Taking-Over Certificate for:

(a) the Works, the Contractor shall include the first half of the Retention Money in a Statement; or
(b) for a Section, the Contractor shall include the relevant percentage of the first half of the Retention Money in a Statement.

This first paragraph is awkwardly worded and is unhelpful for simple implementation by the Contractor. Further it again ignores the opportunity for the Employer to take over a part of the Works.

The paragraph should be amended as follows:

> *"After the Taking-Over Certificate has been (or is deemed to have been) issued for the Works, the Contractor shall include a Claim for the first half of the retention money in a statement under Sub-Clause 14.3 [Application for Interim Payment]. After a Taking-Over Certificate has been (or is deemed to have been) issued for a Section, the Contractor shall include a Claim for the relevant percentage of the first half of the Retention Money applicable to the Section in a statement under Sub-Clause 14.3 [Application for Interim Payment]. After the Taking-Over Certificate has been (or is deemed to have been) issued for a part, the Contractor shall include a Claim for the relevant percentage of the first half of the retention money applicable to the part in a statement under Sub-Clause 14.3 [Application for Interim Payment]."*

After the latest of the expiry dates of the Defects Notification Periods, the Contractor shall include the second half of the Retention Money in a Statement promptly after such latest date. If a Taking-Over Certificate was (or was deemed to have been) issued for a Section, the Contractor shall include the relevant percentage of the second half of the Retention Money in a Statement promptly after the expiry date of the DNP for the Section.

Again the insertion of a few words would assist the clarity of this paragraph as follows:

> *"After the latest of the expiry dates of the Defects Notification Period for the Works, the Contractor shall include a Claim for the outstanding balance of the Retention Money in a Statement promptly after such latest date. If the Taking-Over Certificate was (or was deemed to have been) issued for a Section or Part, the Contractor shall include a Claim for the relevant percentage of the second half of the Retention Money applicable to the Section in a Statement promptly after the expiry date of the DNP for the Section or Part."*

In the next IPC after the Engineer receives any such Statement, the Engineer shall certify the release of the corresponding amount of Retention Money. However, when certifying any release of Retention Money under Sub-Clause 14.6 [*Issue of IPC*], if any work remains to be executed under Clause 11 [*Defects after Taking Over*] or Clause 12 [*Tests after Completion*], the Engineer shall be entitled to withhold certification of the estimated cost of this work until it has been executed.

The intention behind this paragraph is clear. It is intended to give the Engineer power to estimate the costs of any outstanding work under the two identified Clauses and withhold the value of the estimated costs from payment under the imminent IPC. However, that is not what the paragraph states.

A more explicit instruction would be as follows:

> *"In the next IPC after the Engineer received any such Statement, the Engineer may certify the release of the corresponding amount of Retention Money. However, when certifying any release of Retention Money under Sub-Clause 14.6 [Issue of IPC], if any work remains to be executed under Clause 11 [Defects after Taking Over] or Clause 12 [Tests after Completion], the Engineer shall be entitled to estimate the cost of the outstanding work and withhold the value of that cost from the IPC until it has been executed."*

The mandatory direction to the Engineer to release the retention is being removed to give the Engineer discretion.

The relevant percentage for each Section shall be the percentage value of the Section as stated in the Contract Data. If the percentage value of a Section is not stated in the Contract Data, no percentage of either half of the Retention Money shall be released under this Sub-Clause in respect of such Section.

This is an extraordinary contradiction. What is intended is that where the Contract Data is silent then a release of Retention applicable to the Section shall not be made until the release of Retention for the Works is authorised. At that time the Retention for the Section will be included in that final Claim.

This paragraph requires clarification as follows:

> *"The relevant percentage proportion of the Retention Money for each Section will be identified in the Contract Data and/or other relevant document. If the percentage proportion of the Retention Money for each Section is not stated in the Contract Data or and any other document then none of the Retention Money applicable to that Section will be released until the Completion of the Works and Retention for the Section will be included in the Claim made for the release of the Retention of the Works. The same direction applies to a Part of the Works."*

There is no reference to the situation where a "Part" of the Works has been taken over. There will be situations where a "Part" is of particular value should this Sub-Clause should apply. Finally, Contractors should ensure from the Contract Data whether or not there is an allowance for the release of Retention for Sections.

14.10 Statement at Completion

Within 84 days after the Date of Completion of the Works, the Contractor shall submit to the Engineer a Statement at Completion with supporting documents, in accordance with Sub-Clause 14.3 [*Application for Interim Payment*], showing:

(a) the value of all work done in accordance with the Contract up to the Date of Completion of the Works;
(b) any further sums which the Contractor considers to be due at the Date of Completion of the Works; and

(c) an estimate of any other amounts which the Contractor considers have or will become due after the Date of Completion of the Works, under the Contract or otherwise. These estimated amounts shall be shown separately (to those of sub-paragraphs (a) and (b) above) and shall include estimated amounts for:
   (i) Claims for which the Contractor has submitted a Notice under Sub-Clause 20.2 [*Claims For Payment and/or EOT*];
   (ii) any matter referred to the DAAB under Sub-Clause 21.4 [*Obtaining DAAB's Decision*]; and
   (iii) any matter for which a NOD has been given under Sub-Clause 21.4 [*Obtaining DAAB's Decision*].

The Engineer shall then issue an IPC in accordance with Sub-Clause 14.6 [*Issue of IPC*].

This Sub-Clause provides very clear instruction for a Contractor to undertake. First he must identify the value of all the Works carried out up to the date of the Taking-Over Certificate for the Works. Then he must add the value of all other outstanding sums due to him, which do not fall under the categories outlined subsequently. And finally he must list the sums Claimed under the various headings set out in the sub-paragraphs within sub-paragraph (c). The period of 84 days gives the Contractor sufficient time to consider his various heads of claim and make adequate estimates for the same. There is also a very useful provision, which permits the Contractor to consider Claims arising not only out of the Contract but also independently of the Contract.

14.11 Final Statement

Submission by the Contractor of any Statement under the following provisions of this Sub-Clause shall not be delayed by reason of any referral under Sub-Clause 21.4 [*Obtaining DAAB's Decision*] or any arbitration under Sub-Clause 21.6 [*Arbitration*].

Although this paragraph within the Sub-Clause 14.11.1 appears complex it is not. It requires the Contractor to use the same format as previous statements, provide the document in a similar manner as previously provided, and simply replicate the matters set out in Sub-Clause 14.10 sub-paragraphs (a)–(c) inclusive.

14.11.1 Draft Final Statement

Within 56 days after the issue of the Performance Certificate the Contractor shall submit to the Engineer a Draft Final Statement.

This Statement shall:

(a) be in the same form as Statements previously submitted under Sub-Clause 14.3 [*Application for Interim Payment*];
(b) be submitted in one paper-original, one electronic copy and additional paper copies (if any) as stated in the Contract Data; and
(c) show in detail, with supporting documents:
    (i) the value of all work done in accordance with the Contract;
    (ii) any further sums which the Contractor considers to be due at the date of the issue of the Performance Certificate, under the Contract or otherwise; and
    (iii) an estimate of any other amounts which the Contractor considers have or will become due after the issue of the Performance Certificate, under the Contract or otherwise, including estimated amounts by reference to the matters described in sub-paragraphs (c) (i) to (iii) of Sub-Clause 14.10 [*Statement at Completion*]. These estimated amounts shall be shown separately (to those of sub-paragraphs (i) and (ii) above).

Except for any amount under sub-paragraph (iii) above, if the Engineer disagrees with or cannot verify any part of the draft final Statement, the Engineer shall promptly give a Notice to the Contractor. The Contractor shall then submit such further information as the Engineer may reasonably require within the time stated in this Notice, and shall make such changes in the draft as may be agreed between them.

The intent of this paragraph is clear. Where the Engineer has a query as to an amount Claimed in the draft final statement then the Engineer can seek further information. The Contractor shall then give that information. However, the final part of the sentence referring to changes in the draft as may be agreed between them is somewhat unstructured. Therefore, the changes proposed are as follows:

> *"Except for any amount under sub-paragraph (iii) above, if the Engineer disagrees with, or cannot verify, any Claim within the Draft Final Statement, he shall promptly give a Notice to the Contractor identifying the Claim and seeking further information. The Contractor shall then submit such further information as the Engineer may reasonably require within the time stated in this Notice. Subsequently, if required upon review of the information, the Engineer may make such changes to the draft as the Engineer may determine."*

This direction, while useful in intent, being that payment of the Final Statement shall not be hindered or delayed by reason of a referral under Sub-Clauses 21.4 and/or 21.6, however, lacks guidance for the Engineer as to the method of appraising the value of the Final Statement in those particular circumstances.

There is an option for the Engineer to request further information to assist the Engineer in determining the value of particular elements of the claim. Two comments. The first refers to the reference to "This Notice" in the penultimate sentence. Generally, these Conditions avoid any such reference to "this" limiting the wording to "The Notice". That usage is to be preferred. Secondly the Contractor is required to make "such changes" to the Statement as may be agreed. There is no procedure for such agreement process. Also, the use of "such" implies an earlier reference to Changes, which is lacking here.

14.11.2 Agreed Final Statement

If there are no amounts under sub-paragraph (iii) of Sub-Clause 14.11.1 [*Draft Final Statement*], the Contractor shall then prepare and submit to the Engineer the final Statement as agreed (the "Final Statement" in these Conditions).

However, if:

(a) there are amounts under sub-paragraph (iii) of Sub-Clause 14.11.1 [*Draft Final Statement*]; and/or
(b) following discussions between the Engineer and the Contractor, it becomes evident that they cannot agree any amount(s) in the draft final Statement, the Contractor shall then prepare and submit to the Engineer a Statement identifying separately: the agreed amounts,

the estimated amounts and the disagreed amount(s) (the "Partially Agreed Final Statement" in these Conditions).

The heading is misleading. This Sub-Clause contains references to a Final Statement and to a Partially Agreed Final Statement.

Further this Sub-Clause does not take account of monies due under Sub-paragraph (ii) of Sub-Clause 14.11.1.

Therefore, there needs to be a reconsideration of this Sub-Clause as follows:

"Partially Agreed Final Statement and Final Statement

If there are no amounts under sub-paragraphs (ii) and (iii) of Sub-Clause 14.11.1 [Draft Final Statement], the Contractor shall then prepare and submit to the Engineer the Final Statement as agreed (the Final Statement in these Conditions).

However, if: there are amounts under sub-paragraphs (ii) and (iii) under Sub-Clause 14.11.1 [Draft Final Statement]; and/or

following discussions between the Engineer and the Contractor and the making of any changes to the Draft Final Statement which are agreed it becomes evident that they cannot agree certain amount(s) in the Draft Final Statement,

the Contractor shall then prepare and submit to the Engineer a Statement identifying separately:

the agreed Amounts, the estimated Amounts and the disagreed Amounts (the 'Partially Agreed Final Statement' in these Conditions)."

This revision has now explained the procedures and included the amounts outstanding under sub-paragraph (ii) of Sub-Clause 14.11.1 [*Draft Final Statement*], which is omitted in this Sub-Clause.

14.12 Discharge

When submitting the Final Statement or the Partially Agreed Final Statement (as the case may be), the Contractor shall submit a discharge which confirms that the total of such Statement represents full and final settlement of all moneys due to the Contractor under or in connection with the Contract. This discharge may state that the total of the Statement is subject to any payment that may become due in respect of any Dispute for which a DAAB proceeding or arbitration is in progress under Sub-Clause 21.6 [*Arbitration*] and/or that it becomes effective after the Contractor has received:

(a) full payment of the amount certified in the FPC; and
(b) the Performance Security.

If the Contractor fails to submit this discharge, the discharge shall be deemed to have been submitted and to have become effective when the conditions of sub-paragraphs (a) and (b) have been fulfilled.

A discharge under this Sub-Clause shall not affect either Party's liability or entitlement in respect of any Dispute for which a DAAB proceeding or arbitration is in progress under Clause 21 [*Disputes and Arbitration*].

The Contractor would be very unwise to submit a final discharge when there are still amounts outstanding whether they have been referred to a DAAB and/or arbitration or not.

It is also not clear from these general Conditions of Contract how the Parties should proceed in the event that a Partially Agreed Final Statement has been issued. There is further mention of a Partially Agreed Final Statement in Sub-Clause 14.13. However, whether there is an adequate link between the situation where there are disagreements between the Engineer and the Contractor outstanding and the issuance of a FPC will be noted.

14.13 Issue of FPC

Within 28 days after receiving the Final Statement or the Partially Agreed Final Statement (as the case may be), and the discharge under Sub-Clause 14.12 [*Discharge*], the Engineer shall issue to the Employer (with a copy to the Contractor), the Final Payment Certificate, which shall state:

Here are listed the elements which must be taken into account when determining the Final Payment Certificate.

(a) the amount which the Engineer fairly considers is finally due, including any additions and/or deductions which have become due under Sub-Clause 3.7 [*Agreement or Determination*] or under the Contract or otherwise; and

(b) after giving credit to the Employer for all amounts previously paid by the Employer and for all sums to which the Employer is entitled, and after giving credit to the Contractor for all amounts (if any) previously paid by the Contractor and/or received by the Employer under

> the Performance Security, the balance (if any) due from the Employer to the Contractor or from the Contractor to the Employer, as the case may be.

This direction makes clear to the Engineer the various additions and deductions that the Engineer must take into account when establishing the FPC.

> If the Contractor has not submitted a Draft Final Statement within the time specified under Sub-Clause 14.11.1 [*Draft Final Statement*], the Engineer shall request the Contractor to do so. Thereafter, if the Contractor fails to submit a draft final Statement within a period of 28 days, the Engineer shall issue the FPC for such amount as the Engineer fairly considers to be due.

Here it appears that the list set out in the earlier paragraph does not apply. The evaluative process is limited to an amount that the Engineer considers due.

> If:
>
> (i) the Contractor has submitted a Partially Agreed Final Statement under Sub-Clause 14.11.2 [*Agreed Final Statement*]; or
> (ii) no Partially Agreed Final Statement has been submitted by the Contractor but, to the extent that a draft final Statement submitted by the Contractor is deemed by the Engineer to be a Partially Agreed Final Statement the Engineer shall proceed in accordance with Sub-Clause 14.6 [*Issue of IPC*] to issue an IPC.

This paragraph should commence this Sub-Clause as it provides an interim position prior to the issue of an FPC where there are still more matters which both Parties seemingly consider are capable of agreement.

14.14 Cessation of Employer's Liability

> The Employer shall not be liable to the Contractor for any matter or thing under or in connection with the Contract or execution

of the Works, except to the extent that the Contractor shall have included an amount expressly for it in:

(a) the Final Statement or Partially Agreed Final Statement; and
(b) (except for matters or things arising after the issue of the Taking-Over Certificate for the Works) the Statement under Sub-Clause 14.10 [*Statement at Completion*].

The brevity of this clause has not been helpful here. It should be quite clear in the first sentence exactly what the amount expressly is referring to. Therefore, the direction needs to be amended as follows:

> *"The Employer shall not be liable to the Contractor for any matter or thing arising under or in connection with the Contract or execution of the Works, except to the extent that the Contractor shall have expressly included an amount for each such matter or thing in ... the Statement:"*

There is now a connection between the matter or thing and the Contract and the execution of the Works by the use of the word "arising". And the Contractor is required to identify and link each amount with each such matter or thing as referred to earlier.

Unless the Contractor makes or has made a Claim under Sub-Clause 20.2 [*Claims For Payment and/or EOT*] in respect of an amount or amounts under the FPC within 56 days of receiving a copy of the FPC, the Contractor shall be deemed to have accepted the amounts so certified. The Employer shall then have no further liability to the Contractor, other than to pay the amount due under the FPC and return the Performance Security to the Contractor.

This paragraph would benefit from further clarity. It has already been established that the Final Statement or the Partially Agreed Final Statement can contain within it matters arising out of the Contract or in connection with the Contract which the Contractor has expressly reserved. These matters then need to be continuously noted until resolved and the implications of the deeming provision should not come into force until the matters or things identified above have been resolved.

The proposed change therefore be as follows:

> *"Unless the Contractor:*
> *(i) Submits a Claim under Sub-Clause 20.2 [Claims for Payment and all EOT] in respect of the amounts relating to the matters or things listed as required under of*

the first paragraph above contained in the FPC or the partially agreed final statement; ... "

However, this Sub-Clause shall not limit the Employer's liability under the Employer's indemnification obligations, or the Employer's liability in any case of fraud, gross negligence, deliberate default or reckless misconduct by the Employer.

The punctuation is in the wrong place. What is intended is that both the indemnification obligations and the liability shall not be limited in the event of fraud, deliberate default or reckless misconduct.

The direction should be rephrased as follows:

> *"However, this Sub-Clause shall not limit the Employer's liability under his indemnification obligations or his liability in any case of fraud, deliberate default or reckless misconduct by the Employer or the Employer's Personnel."*

14.15 Currencies of Payment

The Contract Price shall be paid in the currency or currencies named in the Contract Data. If more than one currency is so named, payments shall be made as follows:

(a) if the Accepted Contract Amount was expressed in Local Currency only or in Foreign Currency only:

(i) the proportions or amounts of the Local and Foreign Currencies, and the fixed rates of exchange to be used for calculating the payments, shall be as stated in the Contract Data, except as otherwise agreed by both Parties;

(ii) payments and deductions under Sub-Clause 13.4 [*Provisional Sums*] and Sub-Clause 13.6 [*Adjustments for Changes in Laws*] shall be made in the applicable currencies and proportions; and

(iii) other payments and deductions under sub-paragraphs (i) to (iv) of Sub-Clause 14.3 [*Application for Interim Payment*] shall be made in the currencies and proportions specified in sub-paragraph (a)(i) above;

(b) whenever an adjustment is agreed or determined under Sub-Clause 13.2 [*Value Engineering*] or Sub-Clause 13.3 [*Variation Procedure*], the amount payable in each of the applicable currencies shall be specified. For this purpose, reference shall be made to the actual or expected currency proportions of the Cost of the varied work, and to the proportions of various currencies specified in sub-paragraph (a)(i) above;
(c) payment of Delay Damages shall be made in the currencies and proportions specified in the Contract Data;
(d) payment of Performance Damages shall be made in the currencies and proportions specified in the Schedule of Performance Guarantees;
(e) other payments to the Employer by the Contractor shall be made in the currency in which the sum was expended by the Employer, or in such currency as may be agreed by both Parties;
(f) if any amount payable by the Contractor to the Employer in a particular currency exceeds the sum payable by the Employer to the Contractor in that currency, the Employer may recover the balance of this amount from the sums otherwise payable to the Contractor in other currencies; and
(g) if no rates of exchange are stated in the Contract Data, they shall be those prevailing on the Base Date and published by the central bank of the Country.

The first point to note is that there are regular references to the Contract Data. It may well be that the information is missing from the Contract Data but available in the Employer's Requirements or in the Contract itself. As always, these Conditions of Contract should be amended to allow for flexibility.

The first Sub-paragraph makes no sense when read with the introduction in line (a). Line (a) accepts the concept that a single Currency shall be identified. Therefore, the balance of the sub-paragraphs makes no sense in conjunction with line (a).

What is clearly intended is that there shall be a second option describing the Accepted Contract Amount being expressed in a variety of Currencies. This Sub-Clause should be redrafted as follows:

> *"The Contract Price shall be paid in the Currency or Currency is named in the Employer's Requirements, Contract Data Contract or other relevant document.*
>
> *(a) if in the Accepted Contract Amount only one Currency is named then all payments shall be made in that Currency.*
> *(b) if the Accepted Contract Amount is expressed in a combination of a local Currency and a foreign Currency or Currencies then:*
> *(i) the proportions or amounts of the local and foreign Currencies and the fixed rates of exchange to be used for calculating the payments, shall be as stated in the Contract Data, except as otherwise agreed by the Parties;*
> *(ii) payments and deductions under Sub-Clause 13.4 [Provisional Sums] and Sub-Clause 13.6 [Adjustment for Changes in Laws] shall be made in the applicable Currencies and proportions shall be as stated in the Contract Data, except as otherwise agreed by the Parties; and*
> *(iii) other payments and deductions under sub-paragraphs (i) to (iv) of Sub-Clause 14.3 [Application for Interim Payment] shall be made in the Currencies and proportions specified in sub-paragraph (a) (i) above;*
> *(c) whenever an adjustment is agreed or determined under Sub-Clause 13.2 [Value Engineering] or Sub-Clause 13.3 [Variation Procedure], the amount payable in each of the applicable Currencies shall be specified. For this purpose, reference shout be made to the actual or expected Currency proportions of the cost of the varied work, and to the proportions of various Currencies specified in sub-paragraph (a)(i) above;*
> *(d) payment of Delay Damages shall be made in the Currencies and proportion specified in the Contract Data;*
> *(e) payment of Non-performance Damages shall be made in the Currencies and proportions specified in the schedule of Performance Guarantees;*
> *(f) other payments to the Employer by the Contractor shall be made in the Currency in which the sum was expended by the Employer, or in such Currency as may be agreed by both Parties;*
> *(g) if any amount payable by the Contractor to the Employer in a particular Currency exceeds the sum payable by the Employer to the Contractor in that Currency, the Employer may recover the balance of this amount from the sums otherwise payable to the Contractor in other Currencies; and*
> *(h) if no rates of exchange are stated in the Contract Data, they shall be those prevailing on the Base Date and published by the central bank of the country."*

CHAPTER 15

Clause 15 Termination by Employer

15.1 Notice to Correct

If the Contractor fails to carry out any obligation under the Contract the Engineer may, by giving a Notice to the Contractor, require the Contractor to make good the failure and to remedy it within a specified time ("Notice to Correct" in these Conditions).

The bracketed section should follow the word Notice in the second line to make sense. It is far simpler to place a comma after "Notice" and then use the phrase "Notice to correct".

The Notice to Correct shall:

(a) describe the Contractor's failure;
(b) state the Sub-Clause and/or provisions of the Contract under which the Contractor has the obligation; and
(c) specify the time within which the Contractor shall remedy the failure, which shall be reasonable, taking due regard of the nature of the failure and the work and/or other action required to remedy it.

After receiving a Notice to Correct the Contractor shall immediately respond by giving a Notice to the Engineer describing the measures the Contractor will take to remedy the failure, and stating the date on which such measures will be commenced in order to comply with the time specified in the Notice to Correct.

The only comment refers to the use of the word "immediately", whereas elsewhere in these general Conditions of Contract the word "promptly" is used. As always it is recommended that any general Conditions of Contract use one word throughout. Otherwise this is a very simple procedure for the Contractor to follow in response to the receipt of the Notice to Correct.

> The time specified in the Notice to Correct shall not imply any extension of the Time for Completion.

15.2 Termination for Contractor's Default

> Termination of the Contract under this Clause shall not prejudice any other rights of the Employer under the Contract or otherwise.

This direction makes very clear that the Employer has to comply with the Notice requirements within this Sub-Clause in order to effect a Termination for default.

15.2.1 Notice

> The Employer shall be entitled to give a Notice (which shall state that it is given under this Sub-Clause 15.2.1) to the Contractor of the Employer's intention to terminate the Contract or, in the case of sub-paragraph (f), (g) or (h) below a Notice of termination, if the Contractor:
>
> (a) fails to comply with:
> (i) a Notice to Correct;
> (ii) a binding agreement, or final and binding determination, under Sub-Clause 3.7 [*Agreement or Determination*]; or
> (iii) a decision of the DAAB under 21.4 [*Obtaining DAAB's Decision*] (whether binding or final and binding)
> and such failure constitutes a material breach of the Contractor's obligations under the Contract;
>
> (b) abandons the Works or otherwise plainly demonstrates an intention not to continue performance of the Contractor's obligations under the Contract;

(c) without reasonable excuse fails to proceed with the Works in accordance with Clause 8 [*Commencement, Delays and Suspension*] or, if there is a maximum amount of Delay Damages stated in the Contract Data, the Contractor's failure to comply with Sub-Clause 8.2 [*Time for Completion*] is such that the Employer would be entitled to Delay Damages that exceed this maximum amount;

(d) without reasonable excuse fails to comply with a Notice of Rejection given by the Engineer under Sub-Clause 7.5 [*Defects and Rejection*] or an Engineer's instruction under Sub-Clause 7.6 [*Remedial Work*], within 28 days after receiving it;

(e) fails to comply with Sub-Clause 4.2 [*Performance Security*];

(f) subcontracts the whole, or any part of, the Works in breach of Sub-Clause 4.4 [*Subcontractors*], or assigns the Contract without the required agreement under Sub-Clause 1.7 [*Assignment*];

(g) becomes bankrupt or insolvent; goes into liquidation, administration, reorganisation, winding-up or dissolution; becomes subject to the appointment of a liquidator, receiver, administrator, manager or trustee; enters into a composition or arrangement with the Contractor's creditors; or any act is done or any event occurs which is analogous to or has a similar effect to any of these acts or events under applicable Laws;

  or if the Contractor is a JV:

  (i) any of these matters apply to a member of the JV, and

  (ii) the other member(s) do not promptly confirm to the Employer that, in accordance with Sub-Clause 1.14(a) [*Joint and Several Liability*], such member's obligations under the Contract shall be fulfilled in accordance with the Contract; or

(h) is found, based on reasonable evidence, to have engaged in corrupt, fraudulent, collusive or coercive practice at any time in relation to the Works or to the Contract.

This direction makes quite clear that there is a two-stage procedure. The first is where the Employer gives a Notice to the Contractor of his intention to terminate the Contract. Only in two instances may an Employer immediately issue a Termination Notice.

Possibly the order of the examples when a Notice of Intention to Terminate may be given could be reordered. At present it appears an insufficiently structured list.

And further is that within (g) there is a completely separate ground for Termination referring to joint-venture matters.

A restructured order would be as follows:

> *"The Employer shall be entitled to give Notice to the Contractor of his intention to terminate the Contract, Notice of Intention to Terminate, or, in the case of sub-paragraphs (g), (h) and (i) below a Notice of Termination, if the Contractor:*
>
> *(a) fails to comply with Sub-Clause 4.2 [Performance Security];*
> *(b) without reasonable excuse fails to proceeds with the Works in accordance with Clause 8 [Commencement, Delays and Suspension] or, if there is a maximum amount of Delay Damages stated in Contract Data, or elsewhere, his failure to comply with Sub-Clause 8.2 [Time for Completion] is such that the Employer's entitlement to Delay Damages exceeds this maximum amount;*
> *(c) subcontracts the whole, or any part of, the Works in breach of Sub-Clause 4.4 [Subcontractors], or assigns the Contract without the required agreement under Sub-Clause 1.7[Assignment];*
> *(d) abandons the Works or otherwise plainly demonstrates the intention not to continue the performance of his obligations under the Contract;*
> *(e) fails to comply with the Notice to Correct;*
> *(f) without Reasonable excuse fails to comply with the Notice of Rejection given by the Engineer under Sub-Clause 7.5 [Defects and Rejection] or an Engineer's Instruction under Sub-Clause 7.6 [Remedial Work], within 28 days (or other period proposed by the Contractor and agreed by the Engineer) after receiving it;*
> *(g) becomes bankrupt or insolvent, goes into liquidation, as a receiving or administration order made against him, compounds with his creditors, carries on business under a receiver, trustee or manager for the benefit of his creditors, or if any action is done or event occurs which (under applicable Laws) has a similar effect to any of these acts or events;*
> *(h) or if the Contractor is, or is part of, an unincorporated JV:*
> *any of these matters apply to a member of the JV, and the other members do not promptly confirm to the Employer that, in accordance with Sub-Clause 1.14[Joint and Several Liability], such members' obligations under the Contract shall be fulfilled in accordance with the Contract; or*
> *(i) is found, based on the balance of probabilities, to have engaged in corrupt, fraudulent, collusive or coercive practice at any time in relation to the Works or where applicable, to the Contract."*

The reordering of the grounds for issuing a Notice of intention to terminate looks at the circumstances as they arise in the timeline of the progress of the Contract. A Contractor needs to be quickly aware as to which of his failed obligations incurs the sanction of a Notice of intent to terminate from the Employer.

Also, the reference to the JV has been given its own separate ground as all or any of the grounds of Termination could apply although emphasis has been implied that it refers only to insolvency or related bankruptcy matters.

15.2.2 Termination

Unless the Contractor remedies the matter described in a Notice given under Sub-Clause 15.2.1 [*Notice*] within 14 days of receiving the Notice, the Employer may by giving a second Notice to the Contractor immediately terminate the Contract. The date of termination shall be the date the Contractor receives this second Notice.

There is a need for two amendments. The first addresses the period of time, which, in certain circumstances, may be extended: see sub-paragraph (d), which already referred to a 28-day period. The second amendment will address the difference between an intention to terminate Notice and the Notice of Termination.

The proposed changes are follows:

> *"Unless the Contractor remedies the matter described in the Notice, Notice of Intention to Terminate, given under Sub-Clause 15.2.1 [Notice] within 14 days, or such time limit as the Parties may agree or the Engineer may direct, after receiving that Notice, the Employer may give a Notice of Termination to the Contractor immediately terminating the Contract. The date of Termination shall be the date the Contractor receives the Notice of Termination."*

The second Notice mentioned in the existing first paragraph of this Sub-Clause is not a second Notice of Intent to Terminate; it is a first and only Notice stating that it is the Notice of Termination. As presently drafted if implemented as stated it will cause fundamental legal problems.

However, in the case of sub-paragraph (f), (g) or (h) of Sub-Clause 15.2.1 [*Notice*], the Employer may by giving a Notice under Sub-Clause 15.2.1 immediately terminate the Contract and the date of termination shall be the date the Contractor receives this Notice.

This second paragraph is not correct based upon the contents of sub-paragraph (ii) of sub-paragraph (g) which makes clear that in the event that the Contractor is an unincorporated JV, other members may promptly confirm to the Employer that the offending member's obligations under the Contract will be fulfilled in accordance with the Contract. Immediate Termination as mentioned in Sub-Clause 15.2.2 would prevent the other members from having the ability to implement this prompt confirmation.

Therefore, this paragraph needs amendment as follows:

> *"However, in the case of sub-paragraph (g) or (h) of Sub-Clause 15.2.1 [Notice], by giving a Notice under Sub-Clause 15.2.1 the Employer may terminate the Contract immediately unless the provisions of sub-paragraphs (g) or (h) make clear that a period of seven days should be allowed for rational response, otherwise the date upon which the Contractor receives this Notice of Termination will be the date of Termination."*

This amendment allows the remaining JV partners time to respond in the manner suggested by the sub-paragraph. It also allows the Contractor when accused of corruption and other related matters to at least have time to raise valid objections to the charges.

15.2.3 After termination

After termination of the Contract under Sub-Clause 15.2.2 [*Termination*], the Contractor shall:

(a) comply immediately with any reasonable instructions included in a Notice given by the Employer under this Sub-Clause:

  (i) for the assignment of any subcontract; and
  (ii) for the protection of life or property or for the safety of the Works;

(b) deliver to the Engineer:

  (i) any Goods required by the Employer,
  (ii) all Contractor's Documents, and
  (iii) all other design documents made by or for the Contractor; and

(c) leave the Site and, if the Contractor does not do so, the Employer shall have the right to expel the Contractor from the Site.

This is not clear as far as the Contractor is concerned. It appears rational that an Employer should be required to issue Notices for all of the examples set out in this Sub-Clause. Further it is not clear whether the Sub-Clause referred to here encompasses the whole of 15.2 or just 15.2.3.

Therefore, there is a proposed amendment as follows:

"15.2.3 After Termination

After the Notice of Termination has become effective and there is Termination of the Contract under Sub-Clause 15.2.2, the Contractor shall upon receipt of a further Notice

(a) comply immediately with any reasonable instructions given by the Employer under this Sub-Clause:

  (i) for the assignment of any Subcontract; and
  (ii) for the protection of life or property or for the safety of the Works;

(b) deliver to the Engineer:

(i) *any goods required by the Employer,*
(ii) *all Contractor's Documents, and*
(iii) *all other Design Documents made by the Contractor, and*

(c) *leave the Site within a reasonable period of time if not a time agreed between Employer and the Contractor, and if he does not do so, the Employer shall have the right to expel the Contractor from the Site."*

15.2.4 Completion of the Works

After termination under this Sub-Clause, the Employer may complete the Works and/or arrange for any other entities to do so. The Employer and/or these entities may then use any Goods, Contractor's Documents and other design documents made by or on behalf of the Contractor to complete the Works.

After such completion of the Works, the Employer shall give another Notice to the Contractor that the Contractor's Equipment and Temporary Works will be released to the Contractor at or near the Site. The Contractor shall then promptly arrange their removal, at the risk and cost of the Contractor. However, if by this time the Contractor has failed to make a payment due to the Employer, these items may be sold (to the extent permitted by applicable Laws) by the Employer in order to recover this payment. Any balance of the proceeds shall then be paid to the Contractor.

In the second paragraph of this Sub-Clause there is a reference to another Notice. This could be confused for a further Notice relevant to Termination. Therefore, an amendment needs to be made as follows:

"After such completion of the Works, the Employer shall give a Notice to the Contractor stating that the Contractor's Equipment and Temporary Works will be released to the Contractor at or near the Site and stating the date upon which such a release will take place. The Contractor shall then promptly arrange for the removal of his Equipment and Temporary Works at the cost and risk of the Contractor. However, if by the date stated in the Notice, the Contractor has failed to make a payment due to the Employer, these items may be sold by public auction or otherwise, to the extent committed by applicable laws, by the Employer in order to recover the value of this payment. Any balance of the proceeds from the sale shall be then paid promptly to the Contractor."

15.3 Valuation after Termination for Contractor's Default

After termination of the Contract under Sub-Clause 15.2 [*Termination for Contractor's Default*], the Engineer shall proceed under Sub-Clause 3.7 [*Agreement or Determination*] to agree or determine the value of the Permanent Works, Goods and Contractor's Documents, and any other sums due to the Contractor for work executed in accordance with the Contract (and, for the purpose of Sub-Clause 3.7.3 [*Time limits*], the date of termination shall be the date of commencement of the time limit for agreement under Sub-Clause 3.7.3).

This valuation shall include any additions and/or deductions, and the balance due (if any), by reference to the matters described in sub-paragraphs (a) and (b) of Sub-Clause 14.13 [*Issue of FPC*].

This valuation shall not include the value of any Contractor's Documents, Materials, Plant and Permanent Works to the extent that they do not comply with the Contract.

This last paragraph is very severe. It appears to be ignoring any procedure for review and objection. Therefore, it is proposed that this is amended as follows:

> *"This valuation shall not include the value of any Contractor's Documents, Materials, Plant and all Permanent Works to the extent that they have been found, under these Conditions of Contract, not to comply with the Contract, Employer's Requirements, Particular Conditions or any other relevant Documents."*

There has to be a rejection of the various items listed in order for quite properly the value attributed to the items to be excluded.

15.4 Payment after Termination for Contractor's Default

The Employer may withhold payment to the Contractor of the amounts agreed or determined under Sub-Clause 15.3 [*Valuation after Termination for Contractor's Default*] until all the costs, losses and damages (if any) described in the following provisions of this Sub-Clause have been established.

Payment after Termination for Contractor's Default

15.4 The Employer may withhold payment to the Contractor of the amounts agreed or determined under Sub-Clause 15.3 [*Valuation after Termination for Contractor's Default*] until all the costs, losses and damages (if any) described The following provisions of this Sub-Clause have been established.

This is common practice within a whole variety of Contracts.

After termination of the Contract under Sub-Clause 15.2 [*Termination for Contractor's Default*], the Employer shall be entitled subject to Sub-Clause 20.2 [*Claims For Payment and/or EOT*] to payment by the Contractor of:

(a) the additional costs of execution of the Works, and all other costs reasonably incurred by the Employer (including costs incurred in clearing, cleaning and reinstating the Site as described under Sub-Clause 11.11 [*Clearance of Site*]), after allowing for any sum due to the Contractor under Sub-Clause 15.3 [*Valuation after Termination for Contractor's Default*];
(b) any losses and damages suffered by the Employer in completing the Works; and
(c) Delay Damages, if the Works or a Section have not been taken over under Sub-Clause 10.1 [*Taking Over the Works and Sections*] and if the date of termination under Sub-Clause 15.2 [*Termination for Contractor's Default*] occurs after the date corresponding to the Time for Completion of the Works or Section (as the case may be). Such Delay Damages shall be paid for every day that has elapsed between these two dates.

The Employer in these circumstances where there is default by the Contractor should be entitled to deduct the various costs identified before, not after, allowing for any sum due to the Contractor under Sub-Clause 15.3 [*Valuation after Termination for Contractor's Default*]. The Employer should be able to set off these costs against any payments due to the Contractor.

This paragraph should be rewritten as follows:

"After Termination of the Contract under Sub-Clause 15.2 [Termination for Contractor's Default], the Employer shall be entitled, subject to Sub-Clause 20.2[Claims for Payment and/or EOT], to payment by the Contractor, before allowing for any sum due to the Contractor under Sub-Clause 15.3[Valuation after Termination for Contractor's Default], of

(a) the additional costs of execution of the Works, and all other costs reasonably incurred by the Employer, including costs incurred in clearing, cleaning and reinstating the Site as described under Sub-Clause 11.11 [Clearance of Site]; and
(b) any losses and damages suffered by the Employer in completing the Works not included in the previous paragraph; and

(c) *Delay Damages, if the Works, Section or a Part have not been taken over under Sub-Clause 10.1 [Taking Over the Works and Sections] and Sub-Clause 10.2 [Taking over of Parts] and if the date of Termination under Sub-Clause 15.2 [Termination for Contractor's Default] occurs after the date corresponding to the time of Completion of the Works, Section or Part (as the case may be). Such Delay Damages shall be paid for every day that has elapsed between the two dates but shall not exceed any total of Delay Damages that has been identified in the Employer's Requirements, Contract Data, Particular Conditions, Contract or any other document."*

15.5 Termination for Employer's Convenience

The Employer shall be entitled to terminate the Contract at any time for the Employer's convenience, by giving a Notice of such termination to the Contractor (which Notice shall state that it is given under this Sub-Clause 15.5).

No reasons are required to be set out in this Notice apparently.

After giving a Notice to terminate under this Sub-Clause, the Employer shall immediately:

(a) have no right to further use any of the Contractor's Documents, which shall be returned to the Contractor, except those for which the Contractor has received payment or for which payment is due under a Payment Certificate;
(b) if Sub-Clause 4.6 [*Co-operation*] applies, have no right to allow the continued use (if any) of any Contractor's Equipment, Temporary Works, access arrangements and/or other of the Contractor's facilities or services; and
(c) make arrangements to return the Performance Security to the Contractor.

The Conditions imposed upon the Employer do not encompass all the benefits made available by the Contractor. It is quite clear that whether or not cooperation applies, Sub-Clause 4.6 [*Cooperation*], the Contractor should not be required to provide any of its Goods in the widest possible sense to be available to the Employer whether on-Site or off-Site. Further there are various Verification and other Systems, which the Employer is required to be put in place. The Employer shall not have the benefit of the systems unless there is adequate recompense for the Contractor.

This paragraph has not been sufficiently considered.
A redrafting of the paragraph is as follows:

> *"After giving a Notice to Terminate under this Sub-Clause, the Employer shall immediately:*
>
> *(a) have no right to further use of any of the Contractor's Documents, except for those for which the Employer has paid or is about to pay under an IPC and whether or not Sub-Clause 4.6 [Cooperation] applies, have no right to be allowed the continued use (if any) of any of the Contractor's Goods, Personnel, Access Arrangements and/or any other of the Contractor's Systems, facilities or services;*
> *(b) return all and/or any Contractor's Documents to the Contractor except those for which the Contractor has received payment or for which payment is due under a Payment Certificate; and*
> *(c) Make arrangements to return promptly the Performance Security to the Contractor."*

Termination under this Sub-Clause shall take effect 28 days after the later of the dates on which the Contractor receives this Notice or the Employer returns the Performance Security. Unless and until the Contractor has received payment of the amount due under Sub-Clause 15.6 [*Valuation after Termination for Employer's Convenience*], the Employer shall not execute (any part of) the Works or arrange for (any part of) the Works to be executed by any other entities.

These series of directions are clear and consistent.

After this termination, the Contractor shall proceed in accordance with Sub-Clause 16.3 [*Contractor's Obligations After Termination*].

15.6 Valuation after Termination for Employer's Convenience

After termination under Sub-Clause 15.5 [*Termination for Employer's Convenience*] the Contractor shall, as soon as practicable, submit detailed supporting particulars (as reasonably required by the Engineer) of:

(a) the value of work done, which shall include:

(i) the matters described in sub-paragraphs (a) to (e) of Sub-Clause 18.5 [*Optional Termination*], and
(ii) any additions and/or deductions, and the balance due (if any), by reference to the matters described in sub-paragraphs (a) and (b) of Sub-Clause 14.13 [*Issue of FPC*]; and

(b) the amount of any loss of profit or other losses and damages suffered by the Contractor as a result of this termination.

The Engineer shall then proceed under Sub-Clause 3.7 [*Agreement or Determination*] to agree or determine the matters described in sub-paragraphs (a) and (b) above (and, for the purpose of Sub-Clause 3.7.3 [*Time limits*], the date the Engineer receives the Contractor's particulars under this Sub-Clause shall be the date of commencement of the time limit for agreement under Sub-Clause 3.7.3).

The Engineer shall issue a Payment Certificate for the amount so agreed or determined, without the need for the Contractor to submit a Statement.

The wording of Sub-Clause 15.7 dictates that an amendment to Sub-Clause 15.6 is required. Sub-Clause 15.7 refers to the Contractor's Submission. However Sub-Clause 15.6 makes no mention of a Submission but rather refers to the fact that the Contractor shall submit detailed particulars.

Therefore, Sub-Clause 15.6 needs to be amended as follows:

> *"After Termination under Sub-Clause 15.5 [Termination for Employer's Convenience] the Contractor shall, as soon as practicable submit a Submission to the Engineer setting out such detailed particulars, as reasonably required by the Engineer, of ..."*

This amendment then provides for a direct linkage between the two sub-paragraphs.

15.7 Payment after Termination for Employer's Convenience

The Employer shall pay the Contractor the amount certified in the Payment Certificate under Sub-Clause 15.6 [*Valuation after Termination for Employer's Convenience*] within 112 days after the Engineer receives the Contractor's submission under that Sub-Clause.

This direction is clear.

CHAPTER 16

Clause 16 Suspension and Termination by Contractor

16.1 Suspension by Contractor

If:

(a) the Engineer fails to certify in accordance with Sub-Clause 14.6 [*Issue of IPC*];
(b) the Employer fails to provide reasonable evidence in accordance with Sub-Clause 2.4 [*Employer's Financial Arrangements*];
(c) the Employer fails to comply with Sub-Clause 14.7 [*Payment*]; or
(d) the Employer fails to comply with:

  (i) a binding agreement, or final and binding determination under Sub-Clause 3.7 [*Agreement or Determination*]; or
  (ii) a decision of the DAAB under 21.4 [*Obtaining DAAB's Decision*] (whether binding or final and binding) and such failure constitutes a material breach of the Employer's obligations under the Contract, the Contractor may, not less than 21 days after giving a Notice to the Employer (which Notice shall state that it is given under this Sub-Clause 16.1), suspend work (or reduce the rate of work) unless and until the Employer has remedied such default.

This action shall not prejudice the Contractor's entitlements to financing charges under Sub-Clause 14.8 [*Delayed Payment*] and to termination under Sub-Clause 16.2 [*Termination by Contractor*].

If the Employer subsequently remedies the default as described in the above Notice before the Contractor gives a Notice of Termination under Sub-Clause 16.2 [*Termination by Contractor*], the Contractor shall resume normal working as soon as is reasonably practicable.

If the Contractor suffers delay and/or incurs Cost as a result of suspending work (or reducing the rate of work) in accordance with this Sub-Clause, the Contractor shall be entitled subject to Sub-Clause 20.2 [*Claims For Payment and/or EOT*] to EOT and/or payment of such Cost Plus Profit.

Sub-Clause 16.1 addresses Contractor's Entitlement to Suspend Work.

It appears clear that the Contractor is required to give only a single notice to the Employer under this paragraph. The Contractor is not required to give a preliminary notice to the Employer and the Contractor will rely on this Sub-Clause to issue a notice under the terms of the Sub-Clause. The reference to reasonable evidence, while compliant with Sub-Clause 2.4 [*Employer's Financial Arrangements*], should be amplified slightly to read:

> *"Reasonable evidence that the Financial Arrangements are compliant with Sub-Clause 2.4. In the event that the Contractor does not accept that the evidence is reasonable then the Contractor shall apply to the Engineer by Notice for a Sub-Clause 3.5 [Determinations]."*

The right given to the Contractor to reduce the rate of work is not qualified by any requirement of reasonableness. This is an arbitrary power given to the Contractor, appropriately where the Employer and/or the Engineer is in breach of any of these four Sub-Clauses as detailed in this Sub-Clause, however, the addition of the requirement of reasonability or reasonableness would give the DAAB the ability to weigh the actions of the Contractor as against the breach of the Sub-Clauses committed by the Employer and/or the Engineer.

Paragraph 2 of Sub-Clause 16.1 says that:

This action shall not prejudice the Contractor's entitlements to financing charges under Sub-Clause 14.8 [*Delayed Payment*] and to termination under Sub-Clause 16.2 [*Termination by Contractor*].

Here, the use of the word "action" might be inappropriate. The Contractor is empowered under paragraph 1 of this Sub-Clause to, "cease work or reduce the rate of work"; thus it might be more appropriate to state:

> *"that the Contractor's suspension or reduction in the rate of work shall not prejudice etc ..."*

Further, since the right to terminate the contract is fundamental to the Contractor, a separate sub-paragraph could be created to address the fact that the Contractor's acts shall not prejudice his entitlement to terminate the contract rather than linking it with the financing charges. Thus:

> *"the Contractor's acts shall not prejudice his entitlement to terminate the Contract."*

Paragraph 3 states:

> If the Employer subsequently remedies the default as described in the above Notice before the Contractor gives a Notice of Termination under Sub-Clause 16.2 [*Termination by Contractor*], the Contractor shall resume normal working as soon as is reasonably practicable.

This new version of this paragraph removes the specific obligation upon the Employer to produce reasonable evidence. The absence of this specific direction indicates that an Employer might state that the Employer has complied with the requirement, even where it has only partially done so and could require the Contractor, by a Notice, to resume working or increase the pace of work. That action by the Employer would prejudice the rights of the Contractor under this Sub-Clause.

Paragraph 4 states:

> If the Contractor suffers delay and/or incurs Cost as a result of suspending work (or reducing the rate of work) in accordance with this Sub-Clause, the Contractor shall be entitled subject to Sub-Clause 20.2 [*Claims For Payment and/or EOT*] to EOT and/or payment of such Cost Plus Profit.

The key point to note in this paragraph is that the sensible requirement to give two separate Notices to the Engineer have been reduced to a single Notice to be issued in order to comply with the requirements of Sub-Clause 20.1 [*Contractor's Claims*].

The Notice in Sub-Clause 20.1 [*Contractor's Claims*] requires the Contractor to describe the event or circumstances giving rise to the claim. The terms of the notice to the Engineer need not be different in content except to say that the notice is given under Sub-Clause 20.1 [*Contractor's Claims*].

The Contractor could, under the present form, claim under Sub-Clause 8.4 [*Extension of Time for Completion*], both under sub-paragraph (b) which is a delay giving entitlement to an extension of time under a Sub-Clause of these conditions and sub-paragraph (e) any delay, impediment or prevention caused by or attributed abroad to the Employer, the Employer's Personnel, or the Employer's other Contractors on the Site.

For the purposes of Sub-Clause 16.1 [*Contractor's Entitlement to Suspend Work*], paragraph 4, the Contractor should refer to both of these Sub-Clauses when making an application for entitlement as required under this Sub-Clause.

It is very interesting to note that former paragraph 5, which stated:

> *"After receiving this notice, the Engineer shall proceed in accordance with Sub-Clause 3.5, [Determinations], to agree or determine these matters."*

has been removed. Therefore, there is no mechanism other than that provided by Sub-Clause 20.1 [*Contractors Claims*] available to the Contractor to obtain a swift resolution of any partial compliance by the Employer. Nor is there available any reference to the DAAB for either formal or informal guidance to the Parties.

Sub-Clause 16.2 begins:

16.2 Termination by Contractor

Termination of the Contract under this Clause shall not prejudice any other rights of the Contractor, under the Contract or otherwise.

16.2.1 Notice

The Contractor shall be entitled to give a Notice (which shall state that it is given under this Sub-Clause 16.2.1) to the Employer of the Contractor's intention to terminate the Contract or, in the case of sub-paragraph (g)(ii), (h), (i) or (j) below, a Notice of termination, if:

(a) the Contractor does not receive the reasonable evidence within 42 days after giving a Notice under Sub-Clause 16.1 [*Suspension by Contractor*] in respect of a failure to comply with Sub-Clause 2.4 [*Employer's Financial Arrangements*];

(b) the Engineer fails, within 56 days after receiving a Statement and supporting documents, to issue the relevant Payment Certificate;
(c) the Contractor does not receive the amount due under any Payment Certificate within 42 days after the expiry of the time stated in Sub-Clause 14.7 [*Payment*];
(d) the Employer fails to comply with:

  (i) a binding agreement, or final and binding determination under Sub-Clause 3.7 [*Agreement or Determination*]; or
  (ii) a decision of the DAAB under 21.4 [*Obtaining DAAB's Decision*] (whether binding or final and binding) and such failure constitutes a material breach of the Employer's obligations under the Contract;

(e) the Employer substantially fails to perform, and such failure constitutes a material breach of, the Employer's obligations under the Contract;
(f) the Contractor does not receive a Notice of the Commencement Date under Sub-Clause 8.1 [*Commencement of Works*] within 84 days after receiving the Letter of Acceptance;
(g) the Employer:

  (i) fails to comply with Sub-Clause 1.6 [*Contract Agreement*], or
  (ii) assigns the Contract without the required agreement under Sub-Clause 1.7 [*Assignment*];

(h) a prolonged suspension affects the whole of the Works as described in sub-paragraph (b) of Sub-Clause 8.12 [*Prolonged Suspension*];
(i) the Employer becomes bankrupt or insolvent; goes into liquidation, administration, reorganisation, winding-up or dissolution; becomes subject to the appointment of a liquidator, receiver, administrator, manager or trustee; enters into a composition or arrangement with the Employer's creditors; or any act is done or any event

occurs which is analogous to or has a similar effect to any of these acts or events under applicable Laws; or

(j) the Employer is found, based on reasonable evidence, to have engaged in corrupt, fraudulent, collusive or coercive practice at any time in relation to the Works or to the Contract.

Sub-Clause 16.2 addresses termination by the Contractor and begins by identifying the various circumstances where the Contractor shall be entitled to terminate the contract. It is noteworthy that the word "shall" is used here as opposed to "may". Arguably the use of the word "shall" makes clear that where there is a breach of any of the circumstances set out in 16.2.1 [Notice] then an act of termination by the Contractor shall not be able to be subsequently criticised by a DAAB upon review.

The only addition that might be appropriate is the use of the phrase "in good faith", thus: "the Contractor shall be entitled to terminate the contract in good faith if:

(a) the Contractor does not receive the reasonable evidence within 42 days after giving a Notice under Sub-Clause 16.1 [*Suspension by Contractor*] in respect of a failure to comply with Sub-Clause 2.4 [*Employer's Financial Arrangements*];

The use of the phrase "reasonable evidence" emphasises the need to amend the reference to evidence in paragraph 3 to "reasonable evidence". However, it is important to note that there are various time limits applicable to this Sub-Clause "entitlement to terminate". Sub-Clause 2.4 [*Employer's Financial Arrangements*] requires the Employer to submit within 28 days after receiving any request from the Contractor "reasonable evidence" that those arrangements have been made. As noted before, Sub-Clause 2.4 [*Employer's Financial Arrangements*] does not identify the need for a Notice; it is merely a request.

In any event there is a 28-day period within which the Employer must submit the reasonable evidence of the Financial Arrangements. Upon expiration of the 28-day period under Sub-Clause 16.1 [*Suspension by Contractor*] the Contractor may, not less than 21 days after the expiry of the 28-day period, give a Notice to the Employer that the work will be suspended or the rate of activity reduced.

The Employer once notified then has a period of 42 days within which to comply with the request for "reasonable evidence". If there is no compliance within this 42-day period, the Contractor shall then be entitled to terminate the Contract.

During this 42-day period the Contractor has suspended work under Sub-Clause 16.1 [*Suspension by Contractor*] or reduced the rate of work.

Thus, it appears that after the expiry of the 28-day period at whatever time thereafter that the Contractor has issued a notice under Sub-Clause 16.1 [*Suspension by Contractor*] there is a total of 63 days that must expire before the Contractor can terminate the Contract.

> (b) the Engineer fails, within 56 days after receiving a Statement and supporting documents, to issue the relevant Payment Certificate;

The second circumstance when a Contractor shall be entitled to terminate the contract is if the Engineer fails within 56 days after receiving a Statement and supporting documents to issue the relevant Payment Certificate.

The circumstances set out in Sub-Clause 16.2 [*Termination by Contractor*] do not refer to the first paragraph of Sub-Clause 16.1 [*Suspension by Contractor*]. In the first paragraph there is a specific reference to Sub-Clause 14.7 [*Payment*]. Sub-Clause 14.7(b) requires the Employer to pay the Contractor the amount certified in each Interim Payment Certificate within 56 days after the Engineer receives the Statement and supporting documents.

However, sub-paragraph (b) in Sub-Clause 16.2 [*Termination by Contractor*] appears to trigger the power after the 56-day period has expired where the Engineer has failed to issue the Payments Certificate. The Employer is under no obligation to comply with Sub-Clause 14.7 [*Payment*]. While it is the Engineer who is in breach, arguably there must be an implied obligation upon the Employer to enquire of the Engineer if the Engineer has failed to issue the relevant Payment Certificate.

However, in these Conditions of Contract there is not an opportunity for the Employer to make representations to the Engineer, or his Representative, as to why the Engineer fails to perform this or any other of his obligations under the Contract.

On the assumption that the Engineer issues a Payment Certificate 21 days after the expiry of the 56-day period set out in Sub-Clause 14.7 [*Payment*], it appears that the Employer is required, under the Contract, although this matter is not addressed, to comply immediately or, to put the matter beyond doubt, within 24 hours.

There appears, therefore, to be a possible confusion with regard to time limits pertaining to the entitlement to suspend or reduce the rate of work and the entitlement to terminate.

To clarify, the explanation provided in the heading to Sub-Clause 16.2 should be expanded to replicate the key element of Sub-Clause 16.1 [*Suspension by Contractor*] as follows:

> *"Contractor's entitlement to terminate the Contract as opposed simply to Termination by Contractor."*

As this matter is of key importance to the integrity of the Contract and could cause the issue as to timing of Certificates to be a matter disputed before a DAAB some clarity should be provided with respect to the various time periods, which arise from failure to comply with Sub-Clauses 14.6 [*Issue the Interim Payment Certificates*] and 14.7 [*Payment*].

(c) the Contractor does not receive the amount due under any Payment Certificate within 42 days after the expiry of the time stated in Sub-Clause 14.7 [*Payment*];

This circumstance must be read with the previous circumstances. First, the Engineer has failed to issue the relevant Payment Certificate within the 56-day period and thereafter a possible timely or out of time Interim Payment Certificate has been issued and the Employer thereafter fails to pay the amount due within 48 days after the expiry of the time stated in Sub-Clause 14.7 [*Payment*]. The time for expiry is 56 days.

The 56-day period, however, refers to the date when the amount certified in each Certificate is due to be paid and the triggering point to make payment is when the Engineer has received the Statement and supporting documents. These triggering points refer to the Interim Payment Certificate.

Therefore, it appears that the Employer has a period of 56 days within which to pay once the Statement and supporting documents have been received and a further 42 days from whenever the Engineer decides to issue the Interim Payment Certificate. Thus, in fact, an Employer could argue that it is not in breach of Sub-Clause 16.2 [*Termination by Contractor*] to pay on an interim Certificate on the 97th day from the date when the Engineer received the Statement and supporting documents.

The fourth circumstance when the Contractor is entitled to terminate is when:

(d) the Employer fails to comply with:
- (i) a binding agreement, or final and binding determination under Sub-Clause 3.7 [*Agreement or Determination*]; or
- (ii) a decision of the DAAB under 21.4 [*Obtaining DAAB's Decision*] (whether binding or final and binding)

and such failure constitutes a material breach of the Employer's obligations under the Contract;

No reference is made to any specific Sub-Clause within these Conditions of Contract to assist any Party as to when "substantial failure to perform" might occur or be deemed to occur. The key obligations include the obligations:

- to provide access to the Site, Sub-Clause 2.1 [*Right of Access to the Site*],
- to ensure that the Employer's Personnel and other Contractors are on Site, Sub-Clause 2.3 [*Employer's Personnel*],
- to co-operate with the Contractor's efforts required of him under Sub-Clause 4.6 [*Co-operation*],
- to take action similar to those which the Contractor is required to take under sub-paragraphs (a), (b) and (c) of Sub-Clause 4.8 [*Safety Procedures*] and under Sub-Clause 4.18 [*Protection of the Environment*], and
- to ensure that the Employer provides reasonable evidence that his Financial Arrangements are in place, Sub-Clause 2.4 [*Employer's Financial Arrangements*].

Whether a breach of Sub-Clause 2.3 [*Employer's Personnel*] would constitute a major obligation is a matter for debate but without the key Employer's Personnel on Site in a timely manner the progress and execution of the Works could be inhibited.

As has been noted, a breach of Sub-Clause 2.4 [*Employer's Financial Arrangements*] has already been specifically provided for under this Sub-Clause so it appears that the substantial failure to perform can refer mainly to a breach of Sub-Clause 2.1 [*Right of Access to the Site*]; possibly Sub-Clause 2.2 [*Permits, Licences and Approvals*] relating to permits, licences and approvals; again possibly with relation to Sub-Clause 2.3 [*Employer's Personnel*] concerning the requirement to incorporate and address matters of safety and protection.

However, the requirement to pay is not encompassed under the obligations imposed upon the Employer under Clause 2. It is right for the Employer to ensure that Financial Arrangements are made and maintained to enable the Employer to pay in accordance with Clause 14 [*Contract Price and Payment*] the amounts that are required. Further, a fundamental and substantial obligation is described in Sub-Clause 14.2 [*Advance Payment*], which addresses the Advance Payment. The substantial failure of any obligation is not identified under the various circumstances that shall entitle the Contractor to terminate under Sub-Clause 16.2. Therefore, presumably this obligation falls within the general terms of the obligations set out in requiring the Employer substantially to perform his obligations under the Contract and a breach thereof entitles the Contractor to terminate the contract.

(e) the Employer substantially fails to perform, and such failure constitutes a material breach of, the Employer's obligations under the Contract;

This is a very unhelpful clause as it depends upon a Party's interpretation of "substantially". The interpretations are very likely to be different.

> (f) the Contractor does not receive a Notice of the Commencement Date under Sub-Clause 8.1 [*Commencement of Works*] within 84 days after receiving the Letter of Acceptance;

This option to terminate arises where the Works have been suspended for more than 84 days and where the Contractor has requested permission to proceed and the request has been refused or ignored. Where the suspension affects the whole of the Works then this option to terminate is effective.

> (g) the Employer:
>
> (i) fails to comply with Sub-Clause 1.6 [*Contract Agreement*], or
>
> (ii) assigns the Contract without the required agreement under Sub-Clause 1.7 [*Assignment*];

Sub-Clause 1.6 [*Contract Agreement*] requires Parties to enter into a Contract Agreement within 28 days after the Contractor receives the Letter of Acceptance unless the Parties agreed otherwise. Further, the agreement shall be based upon the form annexed to the Particular Conditions. Finally, the Employer shall pay stamp duties and similar charges.

There are at least three circumstances where the Employer might be in breach. The first obviously is where the Employer has not entered into the Agreement within the 28-day period after giving the Contractor the Letter of Acceptance. It is highly likely that the Contractor will have already commenced on the Site in these circumstances. It could be advisable for the Contractor to serve an informal notice on the Employer after the 29th day has expired stating that the Contractor notifies the Employer that it is in substantial breach of an obligation under the Contract and the Contractor may terminate at any time thereafter where the Employer continues to fail to comply with Sub-Clause 1.6 [*Contract Agreemen*t].

Whether the Employer would be in breach where the Agreement is not to be based on the form annexed in the Particular Conditions is a matter for discussion. If the Contract Agreement were sufficiently different from the agreement annexed to the Particular Conditions and biased in favour of the Employer, it would be reasonable for the Contractor to deem the breach of Agreement in such a form to be a substantial breach of an Employer's obligation.

Finally, the failure by the Employer to pay the costs associated with entry into the Contract Agreement and an attempt by the Employer to charge those costs to the Contractor whilst in principle a failure deemed a breach of an obligation by the Employer, in most circumstances, the Contractor will be willing to accept such a breach of this obligation, meaning that this is a theoretical possibility without any practical implication.

The second stipulation addressing Sub-Clause 1.7 [*Assignment*] is more clear cut. Sub-Clause 1.7 [*Assignment*] states that neither Party shall assign the whole or any part of the Contract or any benefit or interest therein. That direction is simple and categorical. However, the Sub-Clause then continues:

> *"However, either Party:*
> *may assign the whole of or any part with the prior agreement of the other Party at the sole discretion of such other Party."*

There is no reference to the word "Contract" in this condition. It is presumed that the fundamental breach would be where either Party assigned the whole or part of the Contract without any prior agreement of the other Party. That interpretation would be proper and sensible as the Party who did not agree to the assignment could or would suffer prejudice or severe prejudice. It would therefore be better to amend the paragraph to

> *"However, either Party:*
> *may assign the whole of or any part of the Contract with the prior agreement of the other Party at the sole discretion of such other Party."*

> (h) a prolonged suspension affects the whole of the Works as described in sub-paragraph (b) of Sub-Clause 8.12 [*Prolonged Suspension*];

This simply reminds the Parties of the effects of a Prolonged Suspension.

> (i) the Employer becomes bankrupt or insolvent; goes into liquidation, administration, reorganisation, winding-up or dissolution; becomes subject to the appointment of a liquidator, receiver, administrator, manager or trustee; enters into a composition or arrangement with the Employer's creditors; or any act is done or any event occurs which is analogous to or has a similar effect to any of these acts or events under applicable Laws;

The rights and obligations of the Employer are clear and this sub-paragraph may be implemented immediately if any of these acts are evident. One of the few problems that may exist under this circumstance is obtaining proof that some of the instances are now in existence. Thus, an Employer may be able to compound with his creditors or carry on business under a manager for the benefit of his creditors without the knowledge of the Contractor. Also, it is not certain that under some of these circumstances a prejudice to the Contractor arises at all.

It has become apparent there is no reference to Sub-Clause 18.5 in the schedule of options set out under 16.2.1. Therefore, there requires to be added a further option as follows:

> *"an event which falls under the definition of exceptional risks as described in Sub-Clause 18.1 and where the execution of a step substantially all the Works in progress have been prevented for a continuous period of 84 days;"*

This category is now added to the category where the Employer is bankrupt or likewise inconvenienced or where there has been corruption or related matters and a Notice of Termination may take immediate effect.

> (j) the Employer is found, based on reasonable evidence, to have engaged in corrupt, fraudulent, collusive or coercive practice at any time in relation to the Works or to the Contract.

16.2.2 Termination

> Unless the Employer remedies the matter described in a Notice given under Sub-Clause 16.2.1 [*Notice*] within 14 days of receiving the Notice, the Contractor may by giving a second Notice to the Employer immediately terminate the Contract. The date of Termination shall then be the date the Employer receives this second Notice.
>
> However, in the case of sub-paragraph (g)(ii), (h), (i) or (j) of Sub-Clause 16.2.1 [*Notice*], by giving a Notice under Sub-Clause 16.2.1 the Contractor may terminate the Contract immediately and the date of Termination shall be the date the Employer receives this Notice.
>
> If the Contractor suffers delay and/or incurs Cost during the above period of 14 days, the Contractor shall be entitled subject to Sub-Clause 20.2 [*Claims For Payment and/or EOT*] to EOT and/or payment of such Cost Plus Profit.

An Employer properly advised would challenge any Notice of Terminate in an attempt to give itself time to negotiate or take alternative measures.

Although there is no reference to Sub-Clause 16.4 it would appear that the first rights that attach to the Contractor are those set out in Sub-Clause 16.4 [*Payment on Termination*]. There will also be various rights related to the Contractor's Goods, Plant and Materials. There will also be rights to litigate or arbitrate where appropriate for such civil or common law breaches of contract as may be available.

16.3 Contractor's Obligations after Termination

> After Termination of the Contract under Sub-Clause 15.5 [*Termination for Employer's Convenience*], Sub-Clause 16.2 [*Termination by Contractor*] or Sub-Clause 18.5 [*Optional Termination*], the Contractor shall promptly:
>
> (a) cease all further work, except for such work as may have been instructed by the Engineer for the protection of life or property or for the safety of the Works. If the Contractor incurs Cost as a result of carrying out such instructed work the Contractor shall be entitled subject to Sub-Clause 20.2 [*Claims For Payment and/or EOT*] to be paid such Cost Plus Profit;
> (b) deliver to the Engineer all Contractor's Documents, Plant, Materials and other work for which the Contractor has received payment; and
> (c) remove all other Goods from the Site, except as necessary for safety, and leave the Site.

Sub-Clause 16.3 examines Cessation of Work and Removal of the Contractor's Equipment.

The first point to note is that there are possible conflicts between the requirements for the Contractor to act promptly, as in this Sub-Clause, and the requirement identified in Sub-Clause 15.5 [*Termination for Employer's Convenience*], which states in the first line of the third paragraph:

> The Termination under this Sub-Clause shall take effect 28 days after the later of the dates on which the Contractor received this Notice or the Employer returns the Performance Security.

It is again one of these instances where the use of a word, in this case "promptly", confuses. In reality, if this Sub-Clause [*Cessation of Work and Removal of the Contractor's Equipment*] is compared with Sub-Clause 15.5 [*Termination for Employer's Convenience*] the instruction to the Contractor is to comply with the requirements of Sub-Clause 15.5 [*Termination for Employer's Convenience*] and not to act promptly.

A further point of conflict arises upon examination of Sub-Clause of 18.5 [*Optional Termination*]. In the first paragraph of this Sub-Clause which identifies an optional termination where the Works, or substantially all Works, have been prevented for a continuous period of 84 days or multiple periods that total more than 140 days due to the same notified Force Majeure event then notice may be given by one Party to the other of termination of the Contract.

Under this stipulation, the termination shall take effect seven days after the notice is given, and the Contractor shall proceed in accordance with Sub-Clause 16.3 [*Contractor's Obligations after Termination*].

Yet again it is necessary to remove the word "promptly".

The first direction is to cease all further work. The only exception is granted for such work as may have been instructed by the Engineer for the protection of life or property or for the safety of the Works. It is possible to allow for each such a situation where the termination has arisen under Sub-Clause 18.5 [*Optional Termination*]. Further, there does not appear in any of the Sub-Clauses referred to in Sub-Clause 16.3 any reference to the requirement that an Engineer should, where applicable, give a Notice to the Contractor to carry out such work as is necessary for the protection of life or property or for the safety of the Works.

Sub-Clause 3.5 [*Engineer's Instructions*] examines the opportunity for the Engineer to give Instructions.

The Engineer may issue to the Contractor (at any time) Instructions and additional or modified drawings that may be necessary for the execution of the Works and the remedying of any defects, in accordance with the Contract.

That is the general instruction. Any Instruction from the Engineer is necessary for the execution of the Works or the remedying of any defects. Thus, Instructions given by the Engineer that shall comply or attempt to comply with Sub-Clause 16.3 [*Contractor's Obligations After Termination*] are not authorised. Therefore, any amendments of these Conditions of Contract should widen the scope of the initial paragraph of Sub-Clause 3.5 [*Engineer's Instructions*] by referring to any situation under Sub-Clause 16.3.

16.4 Payment after Termination by Contractor

After termination under Sub-Clause 16.2 [*Termination by Contractor*], the Employer shall promptly:

(a) pay the Contractor in accordance with Sub-Clause 18.5 [*Optional Termination*]; and

(b) subject to the Contractor's compliance with Sub-Clause 20.2 [*Claims For Payment and/or EOT*], pay the Contractor the amount of any loss of profit or other losses and damages suffered by the Contractor as a result of this Termination.

It is very important to note that the requirement to return the Performance Security has been removed. This Sub-Clause only applies where the Termination has taken place as a result of an action by the Contractor. It is accepted that the opportunities given to the Contractor to terminate under Sub-Clause 16.2 [*Termination by Contractor*] are those generated by any failure on the part of the Employer or his surrogate, the Engineer. However, good the intentions contained within the Conditions of Contract, by instructing the Employer to act promptly there will be situations where the Contractor has not acted reasonably.

In the first opportunity given to the Contractor under Sub-Clause 16.2 [*Termination by Contractor*] the Contractor has to be satisfied that he has not received reasonable evidence within 42 days after giving notice in respect of a failure to comply with Sub-Clause 2.4 [*Employer's Financial Arrangements*]. It could well be that the Contractor has behaved unreasonably. The evidence given by the Employer might have met the requirements of the reasonable professional acting independently and impartially. And the matter could well proceed to a DAAB and subsequently on to arbitration. It might well then be unfortunate or even inappropriate for the Employer to comply with the requirements of Sub-Clause 16.4 [*Payment after Termination by Contractor*] in circumstances where the Contractor is attempting to take a wrongful advantage over the Employer.

CHAPTER 17

Clause 17 Care of the Works and Indemnities

17.1 Responsibility for Care of the Works

Unless the Contract is terminated in accordance with these Conditions or otherwise, subject to Sub-Clause 17.2 [*Liability for Care of the Works*] the Contractor shall take full responsibility for the care of the Works, Goods and Contractor's Documents from the Commencement Date until the Date of Completion of the Works, when responsibility for the care of the Works shall pass to the Employer. If a Taking-Over Certificate is issued (or is deemed to be issued) for any Section or Part, responsibility for the care of the Section or Part shall then pass to the Employer.

If the Contract is terminated in accordance with these Conditions or otherwise, the Contractor shall cease to be responsible for the care of the Works from the date of Termination.

After responsibility has accordingly passed to the Employer, the Contractor shall take responsibility for the care of any work, which is outstanding on the Date of Completion, until this outstanding work has been completed.

If any loss or damage occurs to the Works, Goods or Contractor's Documents, during the period when the Contractor is responsible for their care, from any cause whatsoever except as stated in Sub-Clause 17.2 [*Liability for Care of the Works*], the Contractor shall rectify the loss or damage at the Contractor's risk and cost, so that the Works, Goods, or Contractor's Documents (as the case may be) comply with the Contract.

Clause 17 addresses care of the works and indemnities.

Subclause 17.1 examines responsibility for care of the Works.

The first element of paragraph 1 makes clear that the responsibility for the care of the Works is allocated to the Contractor from the Commencement Date until the Date of Completion.

The second paragraph states that where the Contract is terminated then the Contractor shall cease to be responsible for the care of the Works from the date of Termination. That is not correct. The third paragraph reminds the Contractor that even after responsibility for the care of the Works has passed to the Employer the Contractor shall be responsible for the care of any outstanding work. Therefore, the second paragraph should be amended:

> *"Where the Contract is terminated in accordance with these Conditions or otherwise, the Contractor shall cease to be responsible for the care of the Works from the Date of Termination save as described in this Sub-Clause."*

The last paragraph states, in terms, that the Contractor shall be liable to rectify the loss or damage to any of the described items except where the loss is a result of any cause stated in Sub-Clause 17.2 [*Liability for Care of the Works*].

17.2 Liability for Care of the Works

> The Contractor shall be liable for any loss or damage caused by the Contractor to the Works, Goods or Contractor's Documents after the issue of a Taking-Over Certificate. The Contractor shall also be liable for any loss or damage, which occurs after the issue of a Taking-Over Certificate and which arose from an event which occurred before the issue of this Taking-Over Certificate, for which the Contractor was liable.

Sub-Clause 17.2 examines liability for care of the Works. Paragraph 1 is a virtual repetition of the last paragraph Sub-Clause 17.1 responsibility for care of the Works.

17.3 Intellectual and Industrial Property Rights

> In this Sub-Clause, "infringement" means an infringement (or alleged infringement) of any patent, registered design, copyright, trademark, trade name, trade secret or other intellectual or industrial property right relating to the Works; and "claim" means a third party claim (or proceedings pursuing a third party claim) alleging an infringement.

Sub-Clause 17.3 addresses Intellectual and Industrial Property Rights. There is no definition given in the definition section of this series of general conditions of Contract of Intellectual and Industrial Property Rights.

The first paragraph of the Sub-Clause provides a definition of "infringement". There should be a definition of these rights in the definition clause and there should also be a definition of "infringement" not in this clause but in the definition element of the Conditions of Contract.

In fact, the definition of "infringement" is not a definition and as set out not helpful to either Party. It is confusing to add the term "industrial property" within the Sub-Clause. Any reference to intellectual property rights automatically includes a subset, which addresses industrial property.

The Convention establishing the World Intellectual Property Organisation does not define intellectual property. However, the Organisation lists the following matters as being protected by intellectual property rights:

> "Intellectual property refers to creations of the mind: inventions; literary and artistic works; and symbols, names and images used in commerce. Intellectual property is divided into two categories: Industrial Property includes patents for inventions, trademarks, industrial designs and geographical indications. Copyright covers literary works (such as novels, poems and plays), films, music, artistic works (e.g., drawings, paintings, photographs and sculptures) and architectural design. Rights related to copyright include those of performing artists in their performances, producers of phonograms in their recordings, and broadcasters in their radio and television programmes." (www.wipo.int/edocs/pubdocs/en/intproperty/450/wipo_pub_450.pdf)

A preferred amendment to this first paragraph might be as follows:

> *"This Sub-Clause addresses issues arising from a breach otherwise known as an infringement of Intellectual Property Rights relating to or rising out of the Works."*

By phrasing the introduction in this very simple way there is no restriction on what falls to be defined as intellectual property. In addition, the widening of the scope relating to the rights arising out of or relating to the Works is more comprehensive than the current draft.

> Whenever a Party receives a claim but fails to give notice to the other Party of the Claim within 28 days of receiving it, the first Party shall be deemed to have waived any right to indemnity under this Sub-Clause.

The second paragraph of this Sub-Clause is negatively drafted. A better draft would be:

> *"Whenever a Claim arises the Party receiving notice of the claim shall serve a Notice making the other Party aware of the Claim within 28 days of receiving such notification. If the Party notified of the Claim fails to serve such a Notice on the other Party within the 28 day period when the first Party shall be deemed to have waived any right to indemnity from the other Party under this Sub-Clause."*

The Employer shall indemnify and hold the Contractor harmless against and from any claim (including legal fees and expenses) alleging an infringement, which is or was:

(a) an unavoidable result of the Contractor's compliance with the Employer's Requirements and/or any Variation; or
(b) a result of any Works being used by the Employer:

(i) for a purpose other than that indicated by, or reasonably to be inferred from, the Contract, or
(ii) in conjunction with any thing not supplied by the Contractor, unless such use was disclosed to the Contractor before the Base Date or is stated in the Contract.

The third paragraph requires the Employer to indemnify and hold the Contractor harmless against and from any claim alleging an infringement which is or was an unavoidable result of the Contractor's compliance with the Employer's Requirements or a result of the Works being used by the Employer for a purpose other than that indicated or reasonably to be inferred from the Contract or in conjunction with any thing not supplied by the Contractor unless such use was disclosed to the Contractor before the Base Date or is stated in the Contract.

It is difficult to understand the reason for the insertion of the phrase "an unavoidable result". That phrase implies that the Contractor will have been aware of the implications of likelihood of a breach of an intellectual property right in the Employer's Requirements and have drawn that likelihood to the notice of the Employer and that as a result of the Employer's determination to continue with the progress of the Works to Completion such a breach will be an unavoidable result.

This element of this third paragraph does not anticipate such considerations. It would be preferable to omit any reference to "unavoidable" unless more explanation as to how to avoid such a situation is provided.

The second element of this paragraph addresses a situation where any Works being used by the Employer result in an infringement where the purpose used is not that stated in the Contract or to be inferred from the Contract. It would be better to refer to the use stated in the Employer's Requirements as the Contract may not spell out in sufficient detail the variety or otherwise of the uses to which the Works may be put.

Questions are also raised as to the use of the phrase for a purpose other than that "indicated by". This is very vague. The more suitable phrase would be "stated in".

There is obviously a reason for the insertion of the phrase "or reasonably to be inferred from". Unfortunately, there are two procedural steps required here; the first is the requirement to draw inferences and secondly to investigate whether the process is reasonable.

It might be more appropriate to replace "reasonable" with "logical".

The last element of this third paragraph addresses the situation where the Works are being used by the Employer in conjunction with any thing not supplied by the Contractor unless such use was disclosed to the Contractor before the Base Date or is stated in the Contract.

It is unhelpful to identify such a situation where something, which might cause a breach of intellectual property rights, is being apparently wrongfully used by the Employer in the operation of the Works and is only first identified under the heading of risk of infringement of intellectual property rights.

A Contractor should have the opportunity to notify the Employer by Notice that the use of such thing in conjunction with the operation of the Works may or will cause an infringement of an intellectual property right. It may well be that a Contractor will not appreciate the implications but the opportunity should be given somewhere in these Conditions of Contract.

The Contractor shall indemnify and hold the Employer harmless against and from any other Claim (including legal fees and expenses) alleging an infringement, which arises out of or in relation to:

(i) the Contractor's execution of the Works; or
(ii) the use of Contractor's Equipment.

The fourth paragraph of this Sub-Clause imposes a similar obligation upon the Contractor to indemnify and hold the Employer harmless against any Claim which arises out of or in relation to the Contractor's execution of the Works, the use of the Contractor's Equipment or proper use of the Works.

It is possible to envisage situations, which could possibly arise and provide breaches during the Contractor's execution of the Works and or the use of the Contractor's Equipment. However, it is difficult to see how the Contractor shall be required to hold the Employer harmless against any Claim arising out of or in relation to the proper use of the Works where those Works are those described in the Employer's Requirements and or the Contract.

There must be a duty imposed upon the Employer when refining or defining its Requirements to ensure that the Requirements do not result in a breach either partially or wholly of intellectual property rights of another Party. As the condition is currently written it seems unreasonable to impose such a burden upon the Contractor.

If a Party is entitled to be indemnified under this Sub-Clause, the indemnifying Party may (at the indemnifying Party's cost) assume overall responsibility for negotiating the settlement of the claim, and/or any litigation or arbitration, which may arise from it. The other Party shall, at the request and cost of the indemnifying Party, assist in contesting the Claim. This other Party (and the Contractor's Personnel or the Employer's Personnel, as the case may be) shall not make any admission, which might be prejudicial to the indemnifying Party, unless the indemnifying Party failed to promptly assume overall responsibility for the conduct of any negotiations, litigation or arbitration after being requested to do so by the other Party.

The final paragraph of this Sub-Clause authorises a Party, who is required to indemnify the other Party under this Sub-Clause to assume responsibility for negotiating the settlement of the Claim and/or any litigation or arbitration that may arise from it. The indemnifying Party is authorised to request assistance from the other Party where applicable in contesting the Claim. In doing so it must meet any costs which are likely to arise and indemnify the other Party again to save costs and contesting the Claim. This is helpful.

There is then a very important instruction given to both parties that neither Party shall make an admission, which might prejudice the actions being brought or defence being mounted by the indemnifying Party.

The only situation where an admission might be acceptable is where the indemnifying Party failed to promptly assume overall responsibility for the conduct of the proceedings.

This is not sufficiently explicit and requires an additional procedural step. An additional instruction would be as follows:

> *"Where a Party considers the other Party is not excepting responsibility for the infringement in a prompt and effective manner it shall issue a notice to that Party stating that if the other Party does not commence negotiations promptly that the notifying Party shall make an admission."*

This direction then provides the interim step, which is required in order to avoid any debate as to whether the response was reasonable or not by the indemnifying Party.

17.4 Indemnities by Contractor

The Contractor shall indemnify and hold harmless the Employer, the Employer's Personnel, and their respective agents, against and from all third party claims, damages, losses and expenses (including legal fees and expenses) in respect of:

(a) bodily injury, sickness, disease or death of any person whatsoever arising out of or in the course of or by reason of the Contractor's execution of the Works, unless attributable to any negligence, wilful act or breach of the Contract by the Employer, the Employer's Personnel, or any of their respective agents; and

Subclause 17.4 refers to Indemnities by Contractor.

This is the cover-all Sub-Clause intended to protect the Employer and the Employer's Personnel and agents against a variety of claims, which might otherwise be brought against the Employer.

The first element addresses Claims, which arise from physical injury or death of any person whatsoever, that must include general public, arising out of or in the course of or by reason of the Contractor's execution of the Works.

The only exclusion exists where there has been an act of negligence, wilful act or breach of the Contract by the Employer or the Employer's Personnel or their agents. The exclusion, which arises is attributable to any negligence and wilful act, does not read comfortably. What is intended is that the Employer and its team shall be liable where there has been a deliberately negligent act.

The Employer should be able to argue that as a defence it was not aware that the act or omission was negligent. At the moment this argument does not appear to be available. Secondly there is a reference to a wilful act. There is no meaning to this phrase. What is intended presumably is that the employee and its team will be liable where there is a deliberate act of mischief, which causes death or injury.

(b) damage to or loss of any property, real or personal (other than the Works), to the extent that such damage or loss:
  (i) arises out of or in the course of or by reason of the Contractor's execution of the Works, and
  (ii) is attributable to any negligence, wilful act or breach of the Contract by the Contractor, the Contractor's Personnel, their respective agents, or anyone directly or indirectly employed by any of them.

The second element of this Sub-Clause addresses damage or loss of any property real or personal other than the Works itself provided that the loss or damage arises out of or in the course of or by reason of the Contractor's execution of the

Works and, unusually, is attributable to any negligence, wilful act or breach of the Contract by the Contractor or its team.

It is also useful to note that the exclusion provided in the attributions encompasses not only the Contractor and its team but anyone directly or in directly employed by any of them. This width of protection is not extended to the Employer's team. At present the exclusion provided in the first element of this Sub-Clause is offered and limited to the Employer, the Employer's Personnel or any of their respective agents. The exclusion should be widened to include anyone directly or indirectly employed by any of them in a similar vein as that provided for the Contractor's team.

> To the extent, if any that the Contractor is responsible for the Design of part of the Permanent Works under Sub-Clause 4.1 [*Contractor's General Obligations*], and/or any other design under the Contract the Contractor shall indemnify and hold harmless the Employer against all acts, errors or omissions by the Contractor in carrying out the Contractor's design obligations that result in the Works (or Section or Part or major item of Plant, if any), when completed, not being fit for the purpose(s) for which they are intended under Sub-Clause 4.1 [*Contractor's General Obligations*]

The second element of this Sub-Clause is in addition to the first element and demands that the Contractor shall indemnify and hold harmless the Employer against all errors in the Contractor's Design of the Works and other professional services that result in the Works not being fit for the purposes intended in accordance with Sub-Clause 4.1 [*Contractor's General Obligations*] or resulted in any loss and/or damage for the Employer including legal fees and expenses.

The scope of this last element of the direction is very wide and does not even include any reference to the requirement to issue notices by the Employer, which obviously must be necessary.

17.5 Indemnities by Employer

> The Employer shall indemnify and hold harmless the Contractor, the Contractor's Personnel, and their respective agents, against and from all third party claims, damages, losses and expenses (including legal fees and expenses) in respect of:
>
> (a) bodily injury, sickness, disease or death, or loss of or damage to any property other than the Works, which is

attributable to any negligence, wilful act or breach of the Contract by the Employer, the Employer's Personnel, or any of their respective agents; and

(b) damage to or loss of any property, real or personal (other than the Works), to the extent that such damage or loss arises out of any event described under sub-paragraphs (a) to (f) of Sub-Clause 17.2 [*Liability for Care of the Works*].

Subclause 17.5 refers to Indemnities by the Employer.

The main reason is that the Employer does not have any risks, which arise out of the execution of the Works – the risks are the sole responsibility of the Contractor.

However, the first area where the Employer is required to give indemnifications is where is a physical injury or death arises. There is, however, one outstanding exception. That exception is where the Employer is relieved of any liability to loss or damage to the Works, which is attributable to any negligence, wilful act or breach of the Contract by the Employer and its team.

The scope of the direction should be widened to include his respective agents, or anyone directly or indirectly employed by the Employer.

There is no reason in logic or fairness to exclude the wider definition of the Contractor's team in this Sub-Clause.

17.6 Shared Indemnities

The Contractor's liability to indemnify the Employer, under Sub-Clause 17.4 [*Indemnities by Contractor*] and/or under Sub-Clause 17.3 [*Intellectual and Industrial Property Rights*], shall be reduced proportionately to the extent that any event described under sub-paragraphs (a) to (f) of Sub-Clause 17.2 [*Liability for Care of the Works*] may have contributed to the said damage, loss or injury.

Similarly, the Employer's liability to indemnify the Contractor, under Sub-Clause 17.5 [*Indemnities by Employer*] and/or under Sub-Clause 17.3 [*Intellectual and Industrial Property Rights*], shall be reduced proportionately to the extent that any event for which the Contractor is responsible under Sub-Clause 17.1 [*Responsibility for Care of the Works*] may have contributed to the said damage, loss or injury.

Finally, Sub-Clause 17.6 refers to Shared Indemnities.

This Sub-Clause does not really address Shared Indemnities. What it addresses is that where the other Party may have contributed to the damage, loss or injury then the requirement to indemnify shall be reduced on a proportional basis. The title of the Sub-Clause should be "Reduction of Indemnities".

The concept behind this Sub-Clause is rational and endorsed.

CHAPTER 18

Clause 18 Exceptional Events

18.1 Exceptional Events

"Exceptional Event" means an event or circumstance, which:

(i) is beyond a Party's control;
(ii) the Party could not reasonably have provided against before entering into the Contract;
(iii) having arisen, such Party could not reasonably have avoided or overcome; and
(iv) is not substantially attributable to the other Party.

An Exceptional Event may comprise but is not limited to any of the following events or circumstances provided that conditions (i) to (iv) above are satisfied:

(a) war, hostilities (whether war be declared or not), invasion, act of foreign enemies;
(b) rebellion, terrorism, revolution, insurrection, military or usurped power, or civil war;
(c) riot, commotion or disorder by persons other than the Contractor's Personnel and other employees of the Contractor and Subcontractors;
(d) strike or lockout not solely involving the Contractor's Personnel and other employees of the Contractor and Subcontractors;
(e) encountering munitions of war, explosive materials, ionising radiation or contamination by radio-activity, except

as may be attributable to the Contractor's use of such munitions, explosives, radiation or radio-activity; or

(f) natural catastrophes such as earthquake, tsunami, volcanic activity, hurricane or typhoon.

The key conditions must be met. The first event relates to control. The event must be beyond a Party's control. This test is always subjective. The next condition states that the Party could not have reasonably provided for the event. In order to provide for an event, the event must be foreseeable. The next requirement requires that when the event arose the Party could not have reasonably addressed it. Yet again this is a subjective test. Finally, the event must not be substantially attributable to the Party. Again, this test is very subjective and the question is also what is the effect of "substantial attribution".

These risks are standard in many Contracts. There are always questions to be raised as to when terrorism or revolution becomes an exceptional risk. It is somewhat stretching the imagination to anticipate that a strike, which was involving the Contractor's Personnel, became an exceptional risk when others joined the strike were not the contractors personnel. It is assumed that the encounter with Munitions of war refer both to unexploded bombs of some ancient vintage as well as roadside bombs of the current era. Finally, it is quite difficult to determine when a natural Catastrophe is unforeseeable. In certain parts of the world both earthquakes and volcanic activity are foreseeable, just as in different parts of the world are hurricanes or typhoons.

18.2 Notice of an Exceptional Event

If a Party is or will be prevented from performing any obligations under the Contract due to an Exceptional Event (the "affected Party" in this Clause), then the affected Party shall give a Notice to the other Party of such an Exceptional Event, and shall specify the obligations, the performance of which is or will be prevented (the "prevented obligations" in this Clause).

This is a clear direction to the affected Party to give any Notice promptly as soon as that Party becomes aware of the Exceptional Event. It is also very important that the affected Party includes in the Notice not only those Obligations, which that Party considers that he will be unable to carry out but also those Obligations, which may, not will, be affected by the Exceptional Event.

This Notice shall be given within 14 days after the affected Party became aware, or should have become aware, of the Exceptional Event, and the affected Party shall then be excused performance of the prevented obligations from the date such performance is prevented by the Exceptional Event. If this Notice is received by the other Party after this period of 14 days, the affected Party shall be excused performance of the prevented obligations only from the date on which this Notice is received by the other Party.

The second element of this last paragraph addresses the situation where the Notice was served by the affected Party outside the period of 14 days. It is rather unusual to have this fallback clause within the same paragraph in which the affected Party is given a mandatory instruction to serve the Notice within 14 days. It would be sensible to allow an interim stage to be identified as follows:

> *"Where the Exceptional Event occurs but where the affected Party does not become aware of the effect of the Exceptional Event upon the Obligations until after the 14 day period then the affected Party may serve a Notice upon the other Party identifying the Obligations which are prevented from being carried out together with reasons why the Notice was not served within the 14 day period."*

Thereafter, the affected Party shall be excused performance of the prevented obligations for so long as such Exceptional Event prevents the affected Party from performing them. Other than performance of the prevented obligations, the affected Party shall not be excused performance of all other obligations under the Contract.

However, the obligations of either Party to make payments due to the other Party under the Contract shall not be excused by an Exceptional Event.

The overriding requirement to continue with all other obligations is not as clear as required. Therefore, a revised direction is as follows:

> *"Thereafter, the affected Party shall be excused performance of the prevented obligations for so long as such Exceptional Event prevents him from performing them. Other then performance of the prevented obligations, the affected Party shall not be excused from performing all the other remaining obligations under the Contract."*

18.3 Duty to Minimise Delay

> Each Party shall at all times use all reasonable endeavours to minimise any delay in the performance of the Contract as a result of an Exceptional Event.
>
> If the Exceptional Event has a continuing effect, the affected Party shall give further Notices describing the effect every 28 days after giving the first Notice under Sub-Clause 18.2 [*Notice of an Exceptional Event*].
>
> The affected Party shall immediately give a Notice to the other Party when the affected Party ceases to be affected by the Exceptional Event. If the affected Party fails to do so, the other Party may give a Notice to the affected Party stating that the other Party considers that the affected Party's performance is no longer prevented by the Exceptional Event, with reasons.

This direction to the affected Party sets out an eminently clear procedure for the affected Party to follow. It also provides for a procedure for the unaffected Party to serve what might be called a counter Notice stating that in his opinion the Exceptional Event no longer has any influence on the obligations, which the affected Party has previously relied upon.

18.4 Consequences of an Exceptional Event

> If the Contractor is the affected Party and suffers delay and/or incurs Cost by reason of the Exceptional Event of which he/she gave a Notice under Sub-Clause 18.2 [*Notice of an Exceptional Event*], the Contractor shall be entitled subject to Sub-Clause 20.2 [*Claims For Payment and/or EOT*] to:
>
> (a) EOT; and/or
> (b) if the Exceptional Event is of the kind described in sub-paragraphs (a) to (e) of Sub-Clause 18.1 [*Exceptional Events*] and, in the case of sub-paragraphs (b) to (e) of that Sub-Clause, occurs in the Country, payment of such Cost.

The only concern here for the Contractor is to ensure that the Exceptional Event occurs in the country where the Works are being executed as opposed to elsewhere.

The other important point to note is that this is not a Cost Plus Profit combination; there is merely the provision of cost.

18.5 Optional Termination

If the execution of substantially all the Works in progress is prevented for a continuous period of 84 days by reason of an Exceptional Event of which Notice has been given under Sub-Clause 18.2 [*Notice of an Exceptional Event*], or for multiple periods which total more than 140 days due to the same Exceptional Event, then either Party may give to the other Party a Notice of Termination of the Contract.

The English of this first paragraph is awkward. Questions could be raised as to the meaning of the first sentence. A clearer definition would be as follows:

> *"If the execution of the Works in progress is substantially prevented as to all areas of the Works for a continuous period of 84 days by reason of an Exceptional Event of which Notice has been given under Sub-Clause 18.2 [Notice of an Exceptional Event], or for multiple days which total more than 140 days due to same Exceptional Event, then either Party may give to the other Party a Notice of Termination of the Contract."*

In this event, the date of Termination shall be the date 7 days after the Notice is received by the other Party, and the Contractor shall proceed in accordance with Sub-Clause 16.3 [*Contractor's Obligations After Termination*].

There is no opportunity for the Party receiving the Notice; it will generally be the Employer, to serve a counter Notice with reasons as to why that Contract should not be terminated.

It might be appropriate in future to insert a brief paragraph as follows:

> *"If the Party receiving the Notice does not issue a Notice of Objection within seven days then Termination will become effective."*

After the date of termination the Contractor shall, as soon as practicable, submit detailed supporting particulars (as reasonably required by the Engineer) of the value of the work done, which shall include:

(a) the amounts payable for any work carried out for which a price is stated in the Contract;
(b) the Cost of Plant and Materials ordered for the Works which have been delivered to the Contractor, or of which the Contractor is liable to accept delivery. This Plant and Materials shall become the property of (and be at the risk of) the Employer when paid for by the Employer, and the Contractor shall place the same at the Employer's disposal;
(c) any other Cost or liability which in the circumstances was reasonably incurred by the Contractor in the expectation of completing the Works;
(d) the Cost of removal of Temporary Works and Contractor's Equipment from the Site and the return of these items to the Contractor's place of business in the Contractor's country (or to any other destination(s) at no greater cost); and
(e) the Cost of repatriation of the Contractor's staff and labour employed wholly in connection with the Works at the date of termination.

In some instances these costs will be estimates only and the Contractor should be well aware that budget figures will have to be provided to the Engineer in order to expedite payments.

Sub-paragraph (d) requires clarity as follows:

> *"the cost of removal of temporary Works and Contractor's Equipment from the Site and the return of these items to the Contractor's place of business in his country (or to any other destination(s) at no greater cost). If the Contractor is an unincorporated JV, and the members of the JV are from different countries, this cost shall be determined as at the place of business in the nearest country (to the Site) of such members;"*

The Engineer shall then proceed under Sub-Clause 3.7 [*Agreement or Determination*] to agree or determine the value of work done (and, for the purpose of Sub-Clause 3.7.3 [*Time limits*], the date the Engineer receives the particulars under this Sub-Clause shall be the date of commencement of the time limit for agreement under Sub-Clause 3.7.3).

The Engineer shall issue a Payment Certificate, under Sub-Clause 14.6 [*Issue of IPC*], for the amount so agreed or determined, without the need for the Contractor to submit a Statement.

There is no mention here of the return of the Performance Security or the issue of a Provisional Final Payment Certificate or indeed the Final Payment Certificate.

This is a Termination of the Contract albeit when exceptional circumstances or Events prevent the completion of the Contract. There is every reason why provisions, which apply to a Completion of the Works or a Termination at the option of the Employer should not apply here.

The following should be added to this Sub-Clause:

> *"The Engineer shall after the elapsing of the appropriate period of time, 56 days, Under Sub-Clause 3.7 [Agreement or Determination] issue a Final Payment Certificate.*
>
> *Upon receipt of the FPC the Employer shall promptly return the Performance Security less such sums that have already been paid to the Contractor, if any."*

18.6 Release from Performance under the Law

In addition to any other provision of this Clause, if any event arises outside the control of the Parties (including, but not limited to, an Exceptional Event) which:

(a) makes it impossible or unlawful for either Party or both Parties to fulfil their contractual obligations; or

(b) under the law governing the Contract, entitles the Parties to be released from further performance of the Contract, and if the Parties are unable to agree on an amendment to the Contract that would permit the continued performance of the Contract, then after either Party gives a Notice to the other Party of such event:

(i) the Parties shall be discharged from further performance, and without prejudice to the rights of either Party in respect of any previous breach of the Contract; and

(ii) the amount payable by the Employer to the Contractor shall be the same as would have been payable under Sub-Clause 18.5 [*Optional Termination*], and such amount shall be certified by the Engineer, as if the Contract had been terminated under that Sub-Clause.

The heading here refers to release from performance under the law. This is misleading. In fact, this Clause refers to a supervening event, which makes it impossible for the execution of the Works to continue and this event is outside the control of the Parties. In such a key clause it is thought there should be an opportunity for either Party to refer to the DAAB and/or ICC Arbitration for a confirmatory decision as to whether the Event has the implications given to it by the Party seeking to rely upon it.

CHAPTER 19

Clause 19 Insurance

19.1 General Requirements

Without limiting either Party's obligations or responsibilities under the Contract, the Contractor shall effect and maintain all insurances for which the Contractor is responsible with insurers and in terms, both of which shall be subject to consent by the Employer. These terms shall be consistent with terms (if any) agreed by both Parties before the date of the Letter of Acceptance.

The insurances required to be provided under this Clause are the minimum required by the Employer, and the Contractor may, at the Contractor's own cost, add such other insurances that the Contractor may deem prudent.

This is very important; but this paragraph is without structure. An amendment should be as follows:

> *"Without limiting either Party's obligations or responsibilities under the Contract the Contractor shall affect and maintain or insurances for which he is responsible. The Contractor shall, before placing the insurance cover into effect, submit the details of the insurers and the terms of insurance to the Employer by Notice with a copy to the Engineer. If the Employer does not issue a Notice of Objection within seven days then the Employer will have been deemed to have given consent to both the identified insurers and the terms of the insurance. Where terms of insurance have been agreed by both Parties before the date of the Letter of Acceptance then those terms shall form a proportion of the terms of the required insurance cover."*

Whenever required by the Employer, the Contractor shall produce the insurance policies, which the Contractor is required to effect under the Contract. As each premium is paid, the Contractor shall promptly submit either a copy of each receipt of payment to the Employer (with a copy to the Engineer), or confirmation from the insurers that the premium has been paid.

This Procedure is normal practice.

If the Contractor fails to effect and keep in force any of the insurances required under Sub-Clause 19.2 [*Insurances to be provided by Contractor*] then, and in any such case, the Employer may effect and keep in force such insurances and pay any premium as may be necessary and recover the same from the Contractor from time to time by deducting the amount(s) so paid from any moneys due to the Contractor or otherwise recover the same as a debt from the Contractor. The provisions of Clause 20 [*Employer's and Contractor's Claims*] shall not apply to this Sub-Clause.

In spite of the fact that the right to terminate by reason that the Contractor has failed to effect sufficient or adequate insurance it would be unusual to terminate the Contract for this reason alone where otherwise the progress of the execution of the Works was adequate.

If either the Contractor or the Employer fails to comply with any condition of the insurances effected under the Contract, the Party so failing to comply shall indemnify the other Party against all direct losses and claims (including legal fees and expenses) arising from such failure.

It would be useful to provide for the service of Notices both in procedural terms for this paragraph and the previous paragraph. Such an addition would be as follows:

> *"In the event of a Failure of compliance the Party identifying the failure shall serve a Notice on the other Party informing them of the breach of the obligation."*

This procedural link should be added to both the paragraphs previously identified.

The Contractor shall also be responsible for the following:

(a) notifying the insurers of any changes in the nature, extent or programme for the execution of the Works; and

> (b) the adequacy and validity of the insurances in accordance with the Contract at all times during the performance of the Contract.

This is a serious obligation to impose upon the Contractor. Presumably any breach of this obligation will permit the Employer to implement the Contract under Sub-Clause 15.2 [*Termination for Contractor's Default*].

> The permitted deductible limits allowed in any policy shall not exceed the amounts stated in the Contract Data (if not stated, the amounts agreed with the Employer).

There needs to be an additional sentence here. The addition should be as follows:

> *"At the time when the scope and amount of the Insurance covers are agreed with the Employer the Contractor shall also propose deductible limits to be allowed for each Policy. Within seven days the Employer shall either issue a Notice of No-objection or respond with comments as to why consent should not be given."*

> Where there is a shared liability the loss shall be borne by each Party in proportion to each Party's liability, provided the non-recovery from insurers has not been caused by a breach of this Clause by the Contractor or the Employer. In the event that non-recovery from insurers has been caused by such a breach, the defaulting Party shall bear the loss suffered.

This is a clear definition.

19.2 Insurance to be Provided by the Contractor

> The Contractor shall provide the following insurances:

19.2.1 The Works

> The Contractor shall insure and keep insured in the joint names of the Contractor and the Employer from the Commencement

Date until the date of the issue of the Taking-Over Certificate for the Works:

(a) the Works and Contractor's Documents, together with Materials and Plant for incorporation in the Works, for their full replacement value. The insurance cover shall extend to include loss and damage of any part of the Works as a consequence of failure of elements defectively designed or constructed with defective material or workmanship; and
(b) an additional amount of fifteen percent (15%) of such replacement value (or such other amount as may be specified in the Contract Data) to cover any additional costs incidental to the rectification of loss or damage, including professional fees and the cost of demolition and removal of debris.

The reference to Materials and Plant for incorporation in the Works is unnecessary. The definition of both Materials and separately of Plant make clear that those two definitions only refer to items which form part of the Permanent Works. Secondly where the Design is concerned there are also faults and errors to take into account.

It is also recommended that in the second paragraph the 15% additional cover the cost of demolition and removal of debris and related matters may be too low. Therefore, the first paragraph should be amended as follows:

"The Contractor shall insure, at his own cost, and keep insured in the joint names of the Contractor and the Employer from the Commencement Date until the date of issue of the Taking-Over Certificate for the Works:

(a) The Works and Contractor's Documents, together with the Materials and Plant required, for their replacement value. The insurance cover shall extend to include loss and damage of any part of the Works as a consequence of failure of elements of the Works defectively or erroneously designed or constructed with defective material or as a result of defective workmanship; and
(b) an additional amount of at least 15% of such replacement value (or such other amount as may be specified in the Contract Data, Employer's Requirements, Particular Conditions and/or any other relevant document) to cover additional costs incidental to the rectification on loss or damage, including professional fees and the cost of demolition and removal of the debris."

The insurance cover shall cover the Employer and the Contractor against all loss or damage from whatever cause arising until the issue of the Taking-Over Certificate for the Works. Thereafter, the insurance shall continue until the date of the issue of the Performance Certificate in respect of any incomplete work for loss or damage arising from any cause occurring before the date of the issue of the Taking-Over Certificate for the Works, and for any loss or damage occasioned by the Contractor in the course of any operation carried out by the Contractor for the purpose of complying with the Contractor's obligations under Clause 11 [*Defects after Taking Over*] and Clause 12 [*Tests after Completion*].

It appears that there is a conflict between first and second paragraphs. The first paragraph refers to full replacement value for loss and damage arising as a consequence of defective or erroneous design and/or construction with defective material where there has been defective workmanship whereas the second paragraph refers to loss or damage arising from whatever cause.

It is clear that the first paragraph needs further amendment to reflect the all-encompassing description in the second paragraph. Therefore, sub-paragraph (a) of the first paragraph shall be amended again as follows:

> *"The Works and Contractor's Documents, together with Materials and Plant for their full replacement value. The insurance cover shall extend to cover loss and damage from whatever cause including loss and damage of any part of the Works as a consequence of failure of elements erroneously or defectively designed or constructed with defective material or as a result of defective workmanship; and ..."*

However, the insurance cover provided by the Contractor for the Works may exclude any of the following:

(i) the cost of making good any part of the Works which is defective (including defective material and workmanship) or otherwise does not comply with the Contract, provided that it does not exclude the cost of making good any loss or damage to any other part of the Works attributable to such defect or non-compliance;
(ii) indirect or consequential loss or damage including any reductions in the Contract Price for delay;
(iii) wear and tear, shortages and pilferages; and
(iv) unless otherwise stated in the Contract Data, the risks arising from Exceptional Events.

The various items that may be excluded are rational, with the exception of sub-paragraph (ii) which can, if carried through to its ultimate conclusion, have very serious consequences for the Contractor if not covered by Insurance.

The reference to Contract Data should as always be amended to allow for references elsewhere within the Contract documentation.

The exclusion of the Employer's Risk, which refers to any operation of the forces of nature against which an experienced Contractor could not reasonably have been expected to have taken adequate preventative precautions, is also unwise if Cover is available at a reasonable rate. The cost implications are such that it is recommended that the Contractor raise the matter with the Employer when the matter is not identified in the Contract Data, Employer's Requirements et cetera.

19.2.2 Goods

The Contractor shall insure, in the joint names of the Contractor and the Employer, the Goods and other things brought to Site by the Contractor to the extent specified and/or amount stated in the Contract Data (if not specified or stated, for their full replacement value including delivery to Site).

The Contractor shall maintain this insurance from the time the Goods are delivered to the Site until they are no longer required for the Works.

There is no point in having a reference to Goods and other things brought to Site without some indication of what things are. It is better to widen the definition of equipment than to mention "other things". In fact, the definition of equipment is very wide and encompasses all items of Plant used by the Contractor during the execution of the Works. Therefore, the first paragraph shall be amended as follows:

> *"The Contractor shall insure, in the joint names of the Contractor and the Employer, the Contractor's Goods and Equipment brought to Site by the Contractor for the Contractor's use in executing the Works to the extent specified and/or amount stated in the Contract Data (If not specified or stated, for their full replacement value including delivery to Site)."*

The Contractor shall maintain this insurance from the time the Contractor's Equipment is delivered to the Site until it is no longer required as Contractor's Equipment.

There has to be an addition here to allow for an option for an Engineer to require the specific item to remain on Site. Therefore, this second paragraph should be amended as follows:

> *"The Contractor shall maintain this insurance from the time that the Contractor's Equipment is ready for transportation, during transportation and when it is delivered to the Site until it is removed from Site as no longer necessary for the Contractor's execution of the Works unless the Engineer by Notice requires that the item of Equipment continues to remain on Site."*

19.2.3 Liability for breach of professional duty

To the extent that the Contractor is responsible for the design of part of the Permanent Works under Sub-Clause 4.1 [*Contractor's General Obligations*], and/or any other design under the Contract, and consistent with the indemnities specified in Clause 17 [*Care of the Works and Indemnities*]:

(a) the Contractor shall effect and maintain professional indemnity insurance against liability arising out of any act, error or omission by the Contractor in carrying out the Contractor's Design obligations in an amount not less than that stated in the Contract Data (if not stated, the amount agreed with the Employer); and

A different way of wording this first paragraph is as follows:

> *"The Contractor shall ensure that legal liability of the Contractor arising out of the fault, defect, error or omission, whether negligent or otherwise, of the Contractor or of any person for whom the Contractor is responsible in the carrying out of their professional duties for an amount not less than that stated in the Contract Data, Employer's Requirements et cetera (if not stated, the amount agreed with the Employer within 14 days of receipt of the Letter of Acceptance)."*

(b) if stated in the Contract Data, such professional indemnity insurance shall also indemnify the Contractor against liability arising out of any act, error or omission by the Contractor in carrying out the Contractor's design obligations under the Contract that results in the Works (or Section or Part or major item of Plant, if any), when completed, not being fit for the purpose(s) for which they are intended under Sub-Clause 4.1 [*Contractor's General Obligations*].

What is required is that the cover identified in the first paragraph is explicitly extended so that in the event that the Works are not fit for purpose as a result of a breach of the professional duty on behalf of the Contractor that by reason of the cover the Contractor will be able to put the Employer into the position it would have been had the act or default et cetera not taken place. Therefore, this second paragraph might be altered as follows:

"Such insurance shall contain a specific cover enabling the Contractor to indemnify the Employer for a breach of his professional duty described in the first paragraph which result in the Works not being fit for the purpose(s) intended in accordance with Sub-Clause 4.1 [Contractor's General Obligations], and resulting in any loss or damage to the Employer."

The Contractor shall maintain this insurance for the period specified in the Contract Data.

The last paragraph should be replaced as follows:

"The Contractor shall commence this insurance within 14 days of receipt of the Letter of Acceptance and maintain such insurance for the period specified in the Contract Data, Employer's Requirements or and such period of time which is relevant (if not stated, for a period of 10 years)."

19.2.4 Injury to persons and damage to property

The Contractor shall insure, in the joint names of the Contractor and the Employer, against liabilities for death or injury to any person, or loss of or damage to any property (other than the Works) arising out of the performance of the Contract and occurring before the issue of the Performance Certificate, other than loss or damage caused by an Exceptional Event.

The sole amendment refers to the scope of the incidents when the cover will be effective. Other areas within these Conditions of Contract refer to situations arising out of or in connection with the Contract. The phrase "in connection with" gives a wider and more effective scope.

Therefore, this paragraph should be amended as follows:

"The Contractor shall insure, in the joint names of the Contractor and the Employer, against liability for death or injury to any person, or loss of or damage to any property (other than the Works) arising out of or in connection with the performance of the Contract and occurring before the issuance of the Performance Certificate, other than where the death or injury, loss or damage arises from causes encompassed by the Employer's Risks or an Exceptional Event."

There is an amendment also within the last few words so as to include death and injury as well as loss or damage caused by events falling within the events covered by Employer's risks or an Exceptional Event.

The insurance policy shall include a cross liability clause such that the insurance shall apply to the Contractor and the Employer as separate insureds.

The use of the word "insureds" is unusual in general Contracts of this nature and it is suggested that it should be changed to insured Parties. This cross-liability Clause ensures that risks are shared equally as between the Parties. So that in a situation as here where the Employer and the Contractor are both covered by the same policy that in the event of an injury to a member of the Employer's Personnel caused by a member of the Contractor's Personnel the Employer can rely upon this policy to seek recompense from the Contractor.

Such insurance shall be effected before the Contractor begins any work on the Site and shall remain in force until the issue of the Performance Certificate or a Contract has been terminated and the Contractor has left the Site (whichever is earlier) and shall be for not less than the amount stated in the Contract Data (if not stated, the amount agreed with the Employer).

Using the noun "issue" in this situation is incorrect. Issue usually means dispute or discussion point. The word here required is "issuance". As commented earlier on various occasions there needs to be a simple procedure for agreeing an amount with the Employer. This paragraph therefore needs amendment to incorporate the suggestions as follows:

> *"Such insurance shall be effected before the Contractor begins any work on the Site and shall remain in force until the issuance of the Performance Certificate or the Contract has been terminated and the Contractor has left the Site (which ever is earlier) and shall be for not less than the amount stated in the Contract Data, Employer's Requirements, Particular Conditions or/ and any other relevant Documents (if not stated, the amount agreed with the Employer. The amount shall be that notified to the Contractor by the Employer by a Notice served not less than 14 days before the Letter of Acceptance or at the latest together or forming part of with the Letter of Acceptance)."*

Such insurance shall be effected before the Contractor begins any work on the Site and shall remain in force until the issue of the Performance Certificate and shall be for not less than the amount stated in the Contract Data (if not stated, the amount agreed with the Employer).

This is a sensible direction.

19.2.5 Injury to employees

The Contractor shall effect and maintain insurance against liability for claims, damages, losses and expenses (including legal fees and expenses) arising out of the execution of the Works in respect of injury, sickness, disease or death of any person employed by the Contractor or any of the Contractor's other Personnel.

In these Conditions of Contract it is normal for the insurance to be effected in the joint names of the Contract and the Employer. This policy of joint insurance should continue in Sub-Clause 19.2.5. Also, Sub-Clause 19.2.4 specifically identifies death and injury to any person arising out of or in connection with the performance of the Contract. Personnel, whether of the Employer or the Contractor, are not excluded from this Sub-Clause.

In addition, the heading needs to be amended to encompass death as well as injury. It is also necessary to make clear that the Employer and the Engineer shall maintain insurance against death or injury to the Contractor's and their own staff against an act or neglect of the Employer or the Employer's Personnel. The amended first paragraph shall be as follows:

> *"The Contractor shall effect and maintain insurance in the joint names of the Contractor and the Employer, against liability for Claims, including damages losses and expenses (including legal fees and expenses) arising from injury, sickness, disease order of any person employed by the Contractor or any other member of the Contractor's Personnel or Subcontractors. The Engineer shall be indemnified under this policy of insurance, except that this policy may exclude losses and Claims to the extent that they arise from any act or negligence by the Employer or the Employer's Personnel. The Employer shall affect and maintain insurance against liability for Claims, damages, losses and expenses arising from injury, sickness, disease or death of any person employed by the Contractor, Contractor's Personnel or Subcontractors where the Claim arises from any act or negligence of the Employer or the Employer's Personnel."*

The Employer and the Engineer shall also be indemnified under the policy of insurance, except that this insurance may exclude losses and claims to the extent that they arise from any act or neglect of the Employer or of the Employer's Personnel.

The insurance shall be maintained in full force and effect during the whole time that the Contractor's Personnel are assisting in the execution of the Works. For any person employed by a Subcontractor, the insurance may be effected by the Subcontractor, but the Contractor shall be responsible for the Subcontractor's compliance with this Sub-Clause.

This is not a well-drafted paragraph. The use of the word "assisting" is confusing. A more effectively drafted paragraph is as follows:

> *"The insurance shall be maintained in full force and effect until the issuance of the Performance Certificate and such further period as may be required for the carrying out of the Works required during the Defects Liability Period/DNP. For any person employed by a Subcontractor the insurance shall be effected by the Subcontractor and the Contractor shall be responsible for the Subcontractor's compliance with this Sub-Clause,"*

There is no discretion to be given to the Contractor as to whether or not the Subcontractor is required to effect this insurance. The Requirements is mandatory.

19.2.6 Other insurances required by Laws and by local practice

The Contractor shall provide all other insurances required by the Laws of the countries where (any part of) the Works are being carried out, at the Contractor's own cost.

Other insurances required by local practice (if any) shall be detailed in the Contract Data and the Contractor shall provide such insurances in compliance with the details given, at the Contractor's own cost.

This Sub-Clause is somewhat vague. A more concise direction to the Contractor is as follows:

> *"The Contract Data, Employer's Requirements Particular Conditions or other relevant Documents shall identify the Requirements for Insurance of the Law of the country where the Works are to be executed. If the Contractor executes any part of the work in countries other than that where the Permanent Works are situated then the Contractor must supply to the Employer within 14 days of the Letter of Acceptance a schedule of mandatory Insurance covers required in each country of execution. Within seven days thereafter the Contractor, at his own cost, shall supply to the Employer evidence that the Insurances required have been effected. The Contract Data, Employer's Requirements, Particular Conditions or other relevant document shall identify other Insurances required by local practice. The Contractor shall effect these insurances in compliance with the details given at his own cost."*

CHAPTER 20

Clause 20 Employer's and Contractor's Claims

20.1 Claims

A Claim may arise:

(a) if the Employer considers that the Employer is entitled to any additional payment from the Contractor (or reduction in the Contract Price) and/or to an extension of the DNP;
(b) if the Contractor considers that the Contractor is entitled to any additional payment from the Employer and/or to EOT; or
(c) if either Party considers that he/she is entitled to another entitlement or relief against the other Party. Such other entitlement or relief may be of any kind whatsoever (including in connection with any certificate, determination, instruction, Notice, opinion or valuation of the Engineer) except to the extent that it involves any entitlement referred to in sub-paragraphs (a) and/or (b) above.

It is problematic to understand the use of the word "additional" in conjunction with payment in the first sub-paragraph, sub-paragraph (a).

The concept of this relationship is that the Contractor carries out the execution of the Works and for that execution received payment. The Employer does not receive payment but makes payment. Therefore, it is suggested that the word "additional" is removed. The reference to an extension of the Defects Liability Period is rational.

Sub-paragraph (b) is compatible with the terms of the Conditions of Contract. The reference in sub-paragraph (b) "additional" payment is logical and correct.

The reference in sub-paragraph (c) to another entitlement or relief is confusing. There is no indication as to where is the first entitlement or relief. This reference is related to the use of the word "additional" in the first paragraph; it does not make sense. However, sub-paragraphs (a) and (b) refer to Claims arising in connection with the Contract and omit any reference to arising out of the Contract which is specifically mentioned in sub-paragraph (c).

Even more confusing is the reference in sub-paragraph (c) to the execution of the Works with reference to a relief or entitlement. There is a very wide and explicit reference to entitlements or release arising (this word is missing) in connection with or out of the Contract. The Contract encompasses the execution of the Works so how it is thought that the execution of the Works should be a separate entity is not logical.

Then again in sub-paragraph (c) there are given examples of entitlements or reliefs, which may be in connection with certificates, determination, et cetera. Unfortunately, there are not any references to particular Sub-Clauses within these Conditions of Contract where such powers to grant entitlement or relief are explicitly identified. Further the entitlement for relief does not appear to refer to the Certificate or Notice its self but more inexplicably to something in connection with the Notice or certificate.

In the case of a Claim under sub-paragraph (a) or (b) above, Sub-Clause 20.2 [*Claims For Payment and/or EOT*] shall apply.

This clear linkage between Claims and money or extensions of time for sub-paragraphs (a) and (b) emphasises the uncertainty identified in the procedure described in sub-paragraph (c) where money is apparently not considered as a relief or entitlement under this sub-paragraph.

In the case of a Claim under sub-paragraph (c) above, where the other Party or the Engineer has disagreed with the requested entitlement or relief (or is deemed to have disagreed if he/she does not respond within a reasonable time), a Dispute shall not be deemed to have arisen but the claiming Party may by giving a Notice refer the Claim to the Engineer and Sub-Clause 3.7 [*Agreement or Determination*] shall apply. This Notice shall be given as soon as practicable after the claiming Party becomes aware of the disagreement (or deemed disagreement) and shall include details of the claiming Party's case and the other Party's or the Engineer's disagreement (or deemed disagreement).

These Conditions of Contract are proposing a novel solution to seemingly perceived problems. As the concept seems to be worthy of support it would be helpful to provide a simple procedure for the Claiming Party to follow.

It is proposed to amend this paragraph as follows:

> *"In the case of a Claim made under sub-paragraph (c) above, the Claiming Party may, by giving a Notice, refer the Claim to the Engineer and Sub-Clause 3.7 [Agreement or Determination] shall apply. This Notice shall be given as soon as practicable, and in any event not more than seven days, after the Claiming Party becomes aware that the other Party will not agree to a requested entitlement or relief. The other or requested Party will be deemed to reject a request for an entitlement or relief if in response to such a request that the requested Party has not issued a Notice of No-objection. Any such Notice to the Engineer referred to in the first sentence of this paragraph shall set out the details supporting the request for an entitlement or relief and, where known, set out the reasons for rejection of the request."*

20.2 Claims For Payment and/or EOT

> If either Party considers that he/she is entitled to any additional payment by the other Party (or, in the case of the Employer, a reduction in the Contract Price) and/or to EOT (in the case of the Contractor) or an extension of the DNP (in the case of the Employer) under any Clause of these Conditions or otherwise in connection with the Contract, the following Claim procedure shall apply:

The introduction to Sub-Clause 20.2 should be simplified as follows:

> *"If either Party considers that it may have a Claim under either Sub-Clause 20.1 sub-paragraph (a) or (b) arising out of or in connection with the Contract the following procedure shall apply ...."*

It is worthy of note that Sub-Clause 20.2 omits any reference to be Works.

20.2.1 Notice of Claim

> The claiming Party shall give a Notice to the Engineer, describing the event or circumstance giving rise to the cost, loss, delay or extension of DNP for which the Claim is made as soon as practicable, and no later than 28 days after the claiming Party became aware, or should have become aware, of the event or circumstance (the "Notice of Claim" in these Conditions).

The reference in brackets to the Notice of Claim should be located in the first line of this paragraph. The second point is that it is more practical to give 14 days as the cut-off period after becoming aware of the circumstances or event. The

28-day period can inconvenience the other Party as progress in the execution of the Works can have eliminated all or part of the evidence. It is usual to follow the first Notice with a second Notice within a specified time period setting out greater detail of the event or circumstance. This provision is not allowed for in these Conditions of Contract.

If the claiming Party fails to give a Notice of Claim within this period of 28 days, the claiming Party shall not be entitled to any additional payment, the Contract Price shall not be reduced (in the case of the Employer as the claiming Party), the Time for Completion (in the case of the Contractor as the claiming Party) or the DNP (in the case of the Employer as the claiming Party) shall not be extended, and the other Party shall be discharged from any liability in connection with the event or circumstance giving rise to the Claim.

This paragraph is embarrassingly complex in its structure. A simplified structure is as follows:

> *"If the Claiming Party fails to give a Notice of Claim within the period of 28 days the Claim will fail and the other Party shall be discharged from any liability in connection with the Event or Circumstances giving rise to the Claim."*

20.2.2 Engineer's initial response

If the Engineer considers that the claiming Party has failed to give the Notice of Claim within the period of 28 days under Sub-Clause 20.2.1 [*Notice of Claim*] the Engineer shall, within 14 days after receiving the Notice of Claim, give a Notice to the claiming Party accordingly (with reasons).

This is a modestly draconian Sub-Clause. If this is implemented as described then without question the Claiming Party will be definitely out of time. What should occur is that the Engineer should require the Responding Party to give a Response within seven days and thereafter the Engineer could issue a ruling by a reasoned Notice. What is of considerable concern is that Sub-Clause 20.2.2 does not refer to Sub-Clause 20.3 with its unhelpful heading of "waiver of time limits". Sub-Clause 20.3 is in fact an appeal procedure against the ruling of the Engineer under Sub-Clause 20.2.2.

This Sub-Clause should be amended as follows:

> *"If the Engineer considers that the Claiming Party has failed to give the Notice of Claim within the period of 28 days as required under Sub-Clause 20.2.1 [Notice of Claim]*

he shall, within 14 days after receiving the Notice of Claim give a Notice to the Claiming Party which shall be reasoned stating that the Engineer considers the Notice is out of time. if the Claimant wishes to have the decision of the Engineer reconsidered then the Claimant is referred to the procedure set out in Sub-Clause 20.3 [Waiver of Time Limits]."

The heading to this Sub-Clause is incorrect. The heading should be "Engineer's response as to time limits".

If the Engineer does not give such a Notice within this period of 14 days, the Notice of Claim shall be deemed to be a valid Notice. If the other Party disagrees with such deemed valid Notice of Claim the other Party shall give a Notice to the Engineer, which shall include details of the disagreement. Thereafter, the agreement or determination of the Claim under Sub-Clause 20.2.5 [*Agreement or determination of the Claim*] shall include a review by the Engineer of such disagreement.

If the claiming Party receives a Notice from the Engineer under this Sub-Clause and disagrees with the Engineer or considers there are circumstances which justify late submission of the Notice of Claim, the claiming Party shall include in its fully detailed Claim under Sub-Clause 20.2.4 [*Fully detailed claim*] details of such disagreement or why such late submission is justified (as the case may be).

This direction fails to require the Claiming Party to identify the circumstances, which justify the late submission. The details of the disagreement or why the late submission is justified ignores the requirement to start with the circumstances.

20.2.3 Contemporary records

In this Sub-Clause 20.2, "contemporary records" means records that are prepared or generated at the same time, or immediately after, the event or circumstance giving rise to the Claim.

This definition is a standard definition across a variety of Contracts and emphasises the vital importance of the Claiming Party keeping extensive and detailed records of each event or circumstances, which may, not shall, merit a Claim.

The claiming Party shall keep such contemporary records as may be necessary to substantiate the Claim.

> Without admitting the Employer's liability, the Engineer may monitor the Contractor's contemporary records and/or instruct the Contractor to keep additional contemporary records. The Contractor shall permit the Engineer to inspect all these records during normal working hours (or at other times agreed by the Contractor), and shall if instructed submit copies to the Engineer. Such monitoring, inspection or instruction (if any) by the Engineer shall not imply acceptance of the accuracy or completeness of the Contractor's contemporary records.

This is again a standard procedure across a whole suite of construction Contracts whereby the Contractor is required to keep extensive Records of the circumstance or Event and to make them available at all reasonable times to the Engineer or his Representative. It is not clear why there is a definition of contemporary records in this Sub-Clause. It would be quite rational to put the definition in the definition context at the beginning of the Conditions of Contract. There is not an alternative definition, which could cause a problem.

20.2.4 Fully detailed claim

> In this Sub-Clause 20.2, "fully detailed Claim" means a submission, which includes:
>
> (a) a detailed description of the event or circumstance giving rise to the Claim;
> (b) a statement of the contractual and/or other legal basis of the Claim;
> (c) all contemporary records on which the claiming Party relies; and
> (d) detailed supporting particulars of the amount of additional payment claimed (or amount of reduction of the Contract Price in the case of the Employer as the claiming Party), and/or EOT claimed (in the case of the Contractor) or extension of the DNP claimed (in the case of the Employer).

It will not always be possible for the Claimant to be able to quantify the exact amount of Payment, which should be claimed. There may be many reasons including continuing effect of the Event and Circumstances, which may not be obvious during the period in which the Claim must be made but which will emerge thereafter. At present there is no provision for a provisional figure to be submitted.

> Within either:
>
> (i) 84 days after the claiming Party became aware, or should have become aware, of the event or circumstance giving rise to the Claim, or
> (ii) such other period (if any) as may be proposed by the claiming Party and agreed by the Engineer the claiming Party shall submit to the Engineer a fully detailed Claim.
>
> If the event or circumstance giving rise to the Claim has a continuing effect, Sub-Clause 20.2.6 [*Claims of continuing effect*] shall apply.

It is very important to note that the Claiming Party has only a further 14 days within which to bring together a fully detailed Claim after the expiry of the initial Notice Period.

It is also suggested that the element setting out the time limits should introduce the Sub-Clause and not be located after the Definition.

> If within this time limit the claiming Party fails to submit the statement under sub-paragraph (b) above, the Notice of Claim shall be deemed to have lapsed, it shall no longer be considered as a valid Notice, and the Engineer shall, within 14 days after this time limit has expired, give a Notice to the claiming Party accordingly.

The reason why the Claim should be invalidated should the Contractual basis for the Claim not be submitted within the time limit is not clear. This direction should be amended to allow for the Engineer to reject any Claim where any of the Requirements of a submission are lacking. The alternative solution would be to make clear that the Engineer can consider submissions which do not include the Requirements in full of sub-paragraphs (a), (c) and (d) and thus give some flexibility to the Claimant to add more detail within a certain time period which the Engineer could provide.

If the Engineer does not give such a Notice within this period of 14 days, the Notice of Claim shall be deemed to be a valid Notice. If the other Party disagrees with such deemed valid Notice of Claim the other Party shall give a Notice to the Engineer, which shall include details of the disagreement. Thereafter, the agreement or determination of the Claim under Sub-Clause 20.2.5 [*Agreement or determination of the Claim*] shall include a review by the Engineer of such disagreement.

If the claiming Party receives a Notice from the Engineer under this Sub-Clause 20.2.4 and if the claiming Party disagrees with such Notice or considers there are circumstances that justify late submission of the statement under sub-paragraph (b) above, the fully detailed claim shall include details of the claiming Party's disagreement or why such late submission is justified (as the case may be).

20.2.5 Agreement or determination of the claim

After receiving a fully detailed Claim under Sub-Clause 20.2.4 [*Fully detailed Claim*], or an interim or final fully detailed Claim (as the case may be) under Sub-Clause 20.2.6 [*Claims of continuing effect*], the Engineer shall proceed under Sub-Clause 3.7 [*Agreement or Determination*] to agree or determine:

(a) the additional payment (if any) to which the claiming Party is entitled or the reduction of the Contract Price (in the case of the Employer as the claiming Party); and/or
(b) the extension (if any) of the Time for Completion (before or after its expiry) under Sub-Clause 8.5 [*Extension of Time for Completion*] (in the case of the Contractor as the claiming Party), or the extension (if any) of the DNP (before its expiry) under Sub-Clause 11.3 [*Extension of Defects Notification Period*] (in the case of the Employer as the claiming Party), to which the claiming Party is entitled under the Contract.

As with earlier drafting of this Clause 20 over complexity does not benefit the user. A proposed amendment is as follows:

> *"Except where the Engineer has given a Notice under Sub-Clause 20.2.2 [Engineer's Ruling on Time Limits] or under Sub-Clause 20.2.4 [Fully Detailed Claim], he shall proceed under Sub-Clause 3.7 [Agreement or Determination] to agree or determine:*
>
> *(a) any additional payment to which the Claiming Party is entitled or any reduction in the Contract Price where appropriate; and/or*
> *(b) any Extension Of Time for Completion, before or after its expiry, under Sub-Clause 8.5 [Extension of Time for Completion], where appropriate, or any extension of the DNP, before its expiry under Sub-Clause 11.3 [Extension of Defects Notification Period], where appropriate."*

If the Engineer has given a Notice under Sub-Clause 20.2.2 [*Engineer's initial response*] and/or under Sub-Clause 20.2.4 [*Fully detailed Claim*], the Claim shall nevertheless be agreed or determined in accordance with this Sub-Clause 20.2.5. The agreement or determination of the Claim shall include whether or not the Notice of Claim shall be treated as a valid Notice taking account of the details (if any) included in the fully detailed claim of the claiming Party's disagreement with such Notice(s) or why late submission is justified (as the case may be). The circumstances, which may be taken into account (but shall not be binding) may include:

- whether or to what extent the other Party would be prejudiced by acceptance of the late submission;
- in the case of the time limit under Sub-Clause 20.2.1 [*Notice of Claim*], any evidence of the other Party's prior knowledge of the event or circumstance giving rise to the Claim, which the claiming Party may include in its supporting particulars; and/or
- in the case of the time limit under Sub-Clause 20.2.4 [*Fully detailed Claim*], any evidence of the other Party's prior knowledge of the contractual and/or other legal basis of the Claim, which the claiming Party may include in its supporting particulars.

It is not clear why it is necessary to trigger references to the DAAB at this stage. There is insufficient guidance in this paragraph to either Party. The amended paragraph should be as follows:

> *"Where the Engineer has given a Notice under Sub-Clause 20.2.2 [Ruling upon Time Limits] or under Sub-Clause 20.2.4 [Fully Detailed Claim] the Claiming Party must be aware that the obligation upon the Engineer to proceed to agree or determine the Claim can only proceed should the procedures of Sub-Clause 20.3 [Waiver of Time Limits] be activated."*

The Paragraph as written does not alert the Claiming Party of the very tight time Schedules inherent in general Conditions of Contract.

> If, having received the fully detailed Claim under Sub-Clause 20.2.4 [*Fully detailed Claim*], or in the case of a Claim under Sub-Clause 20.2.6 [*Claims of continuing effect*] an interim or final fully detailed Claim (as the case may be), the Engineer requires necessary additional particulars:
>
> (i) he/she shall promptly give a Notice to the claiming Party, describing the additional particulars and the reasons for requiring them;
> (ii) he/she shall nevertheless give his/her response on the contractual or other legal basis of the Claim, by giving a Notice to the claiming Party, within the time limit for agreement under Sub-Clause 3.7.3 [*Time limits*];
> (iii) as soon as practicable after receiving the Notice under sub-paragraph (i) above, the claiming Party shall submit the additional particulars; and
> (iv) the Engineer shall then proceed under Sub-Clause 3.7 [*Agreement or Determination*] to agree or determine the matters under sub-paragraphs (a) and/or (b) above (and, for the purpose of Sub-Clause 3.7.3 [*Time limits*], the date the Engineer receives the additional particulars from the claiming Party shall be the date of commencement of the time limit for agreement under Sub-Clause 3.7.3).

This paragraph within Sub-Clause 20.2.5 [*Agreement or Determination*] appears, for some reason, to impose unnecessary restrictions upon the Engineer. The Engineer is entitled to seek further particulars with reasons. Those further particulars will assist the Engineer as to determining the Contractual or other basis for the Claim. However, sub-paragraph (ii) of this Sub-Clause states that whatever the status of the additional information sought, the Engineer must give a response as to the validity or otherwise of the basis for the Claim within 42 days.

Upon careful reading of Sub-Clause 3.7.3 [*Time-Limits*] there is a reference to 42 days or such other period as shall be agreed between the Parties.

This very important Sub-Clause needs a little revision in order to assist the Engineer to make the most effective response as to whether the Claim is properly Contractually based or based on another basis.

A proposed revision is as follows:

> *"If, having received the fully detailed Claim under Sub-Clause 20.2.4 [Fully Detailed Claim], or in the case of the Claim under Sub-Clause 20.2.6 [Claims of Continuing Effect] the final fully detailed Claim, the Engineer requires additional particulars:*
>
> *(i) he shall promptly give a Notice to be Claiming Party, describing the additional particulars and his reasons for requiring them and within the Notice the Engineer shall identify the date when the original Notice was served by the claiming Party;*
> *(ii) if the further or additional particulars will affect the response as to the Contractual or other basis of the Claim then the Engineer will not be required to provide a response within the 42 day time limit imposed under Sub-Clause 3.7.3 [Time Limits];*
> *(iii) if however the Engineer is able to give his response on the Contractual or other basis for the Claim without further particulars then he shall do so by giving a Notice limited to the Contractual or other basis for the Claim within the 42 day time limit identified under Sub-Clause 3.7.3 [Time Limits];*
> *(iv) as soon as practicable after receiving the Notice under sub-paragraph (i) above, the claiming Party shall submit the additional particulars; and*
> *(v) The Engineer shall then proceed under Sub-Clause 3.7 [Agreement or Determination] to agree or determine the matters under sub-paragraphs (a) and/or (b) above (and, for the purposes of Sub-Clause 3.7.3 [Time Limits] the date the Engineer received the additional particulars from the Claimant Party shall be the date of the Commencement of the period of 42 days, unless extended by agreement of the Parties, shall apply."*

20.2.6 Claims of continuing effect

> If the event or circumstance giving rise to a Claim under this Sub-Clause 20.2 has a continuing effect:
>
> (a) the fully detailed Claim submitted under Sub-Clause 20.2.4 [*Fully detailed Claim*] shall be considered as interim;
> (b) in respect of this first interim fully detailed Claim, the Engineer shall give his/her response on the contractual

> or other legal basis of the Claim, by giving a Notice to the claiming Party, within the time limit for agreement under Sub-Clause 3.7.3 [*Time limits*];
>
> (c) after submitting the first interim fully detailed Claim the claiming Party shall submit further interim fully detailed Claims at monthly intervals, giving the accumulated amount of additional payment claimed (or the reduction of the Contract Price, in the case of the Employer as the claiming Party), and/or extension of time claimed (in the case of the Contractor as the claiming Party) or extension of the DNP (in the case of the Employer as the claiming Party); and
>
> (d) the claiming Party shall submit a final fully detailed Claim within 28 days after the end of the effects resulting from the event or circumstance, or within such other period as may be proposed by the claiming Party and agreed by the Engineer. This final fully detailed claim shall give the total amount of additional payment claimed (or the reduction of the Contract Price, in the case of the Employer as the claiming Party) and/or extension of time claimed (in the case of the Contractor as the claiming Party) or extension of the DNP (in the case of the Employer as the claiming Party]

With respect to sub-paragraph (b) the same concerns are raised as to the requirement for the Engineer to make a decision as to whether or not the basis for the Claim arises out of the Contract or otherwise as were raised earlier. Subsequently sub-paragraphs (c) and (d) are consistent and logical.

20.2.7 General requirements

> After receiving the Notice of Claim, and until the Claim is agreed or determined under Sub-Clause 20.2.5 [*Agreement or determination of the Claim*], in each Payment Certificate the Engineer shall include such amounts for any Claim as have been reasonably substantiated as due to the claiming Party under the relevant provision of the Contract.

This provision does not take account of an Employer claiming a reduction of the Contract Price. Therefore, this paragraph requires amendment as follows:

> *"After receiving the Notice of Claim, and the Notice of Continuing Effect, and until the Claim is agreed or determined under Sub-Clause 20.2.5 [Agreement or Determination of the Claim], in each Payment Certificate and the Engineer shall include such amounts or deductions, as the case may be, for any Claim that have been reasonably substantiated as due to the Claiming Party under the relevant provision of the Contract."*

> The Employer shall only be entitled to claim any payment from the Contractor and/or to extend the DNP, or set off against or make any deduction from any amount due to the Contractor, by complying with this Sub-Clause 20.2.

This is a general failsafe mechanism to ensure that the Employer does not make unauthorised deductions from the payments due to the Contractor.

> The requirements of this Sub-Clause 20.2 are in addition to those of any other Sub-Clause, which may apply to the Claim. If the claiming Party fails to comply with this or any other Sub-Clause in relation to the Claim, any additional payment and/or any EOT (in the case of the Contractor as the claiming Party) or extension of the DNP (in the case of the Employer as the claiming Party), shall take account of the extent (if any) to which the failure has prevented or prejudiced proper investigation of the Claim by the Engineer.

It is quite clear that these last two paragraphs do not belong in Sub-Clause 20.2.6. These are general directions and are properly heading General Requirements. The ideal location for these paragraphs should be under the heading of sub-paragraph 20.2 and demonstrated as follows:

> *"Where either Party considers himself to be entitled to any additional payment or, as the case may be, reduction in the Contract price, and/or to EOT and more, as the case may be to an extension of the DNP and any Clause of these Conditions or otherwise in connection or arising out of the Contract the Party shall follow the procedures set out in the subsequent paragraphs."*

Then add the following phrase:

> *"The Claiming Party shall apply the following procedure:"*

CHAPTER 21

Clause 21 Disputes and Arbitration

21.1 Constitution of the DAAB

Disputes shall be decided by a DAAB in accordance with Sub-Clause 21.4 [*Obtaining DAAB's Decision*]. The Parties shall jointly appoint the member(s) of the DAAB within the time stated in the Contract Data (if not stated, 28 days) after the date the Contractor receives the Letter of Acceptance.

This paragraph introduces the concept of the DAAB. The second sentence refers to a mandatory duty to appoint the member or members within a certain period of time. Unfortunately, this is a discretionary duty and frequently the Employer does not make any appointment leaving the DAAB process suspended.

The other point to keep in mind is that the Contractor may not always receive a Letter of Acceptance but may be instructed by the Employer to commence the execution of the Works. In that event according to this paragraph of Sub-Clause 21.1 there is no authority to constitute a DAAB.

The DAAB shall comprise, as stated in the Contract Data, either one suitably qualified member (the "sole member") or three suitably qualified members (the "members"). If the number is not so stated, and the Parties do not agree otherwise, the DAAB shall comprise three members.

The paragraph does not address the situation where the Employer's Requirements may identify the numbers and qualifications of the DAAB. It could appear that if a DAAB is constituted based on information provided in the Employer's Requirements it could be challenged.

This paragraph should be amended as follows:

> *"The DAAB shall comprise, as stated in the Contract Data or, as the case maybe, elsewhere, either one suitably qualified member (the sole member) or three suitably qualified members (the members). If the number is not so stated, and the Parties do not agree otherwise, the DAAB shall compromise three members."*

> The sole member or three members (as the case may be) shall be selected from those named in the list in the Contract Data, other than anyone who is unable or unwilling to accept appointment to the DAAB.

This paragraph is unnecessarily restrictive. It appears to make clear that the only location for identifying the members of the DAAB is in a list in the Contract Data. This cannot be right.

An amendment is required as follows:

> *"The sole member or the three members (as the case may be) shall be selected from those named in the list in the Contract Data or in the case that there is no list in the Contract Data then from any other document which contains such a list and where there is no such list from an appointing authority. Where the person nominated to be a member of the DAAB is unwilling or unable to accept the appointment then the appointment shall be invalid."*

> If the DAAB is to comprise three members, each Party shall select one member for the agreement of the other Party. The Parties shall consult both these members and shall agree the third member, who shall be appointed to act as chairperson.

This paragraph is meaningless at the present time. It states that each Party shall select one member for the agreement of the other Party. Agreement as to what? The members of the DAAB are either selected from a list or appointed by an appointing authority. Secondly this paragraph directs the Parties to consult the two members representing them. There is no indication as to what the consulting process concerns.

This paragraph also requires amendment as follows:

> *"If a DAAB is to comprise three members, each Party shall select one member of the DAAB. Thereafter The Parties shall consult the two members of the DAAB as to the identity of the third member and shall use their best efforts to agree upon the identity of the third member who, when appointed, shall be appointed to act as chair of the DAAB. In the event that the Parties are unable to agree upon the identity of a third member then within 14 days of a Party notifying the other Party of the Failure to agree then the notifying Party shall require the appointing authority to make the appointment or as directed in Sub-Clause 21.2 [Failure to appoint DAAB Members]."*

The DAAB shall be deemed to be constituted on the date that the Parties and the sole member or the three members (as the case may be) of the DAAB have all signed a DAAB Agreement.

This requirement for all Parties to sign the agreement is another example where an unwilling or recalcitrant Party can totally stymie the constitution of a DAAB. It is recommended that institutions who provide for such entities as dispute adjudication boards consider greater flexibility in order for such boards to be constituted. There are frequently instances where the Contractor is willing to sign a DAAB agreement taking responsibility the fees of all members of the DAAB.

However, professional opinion is of the view that unless the Employer is willing to support the DAAB process by signing the agreement then to have a unilateral support is of no value.

However, in view of strong views to the contrary a suggested revision of this paragraph is as follows:

> *"The DAAB shall be deemed to be constituted on the date that the Parties and the sole member or the three members (as the case may be) of the DAAB have all signed the DAAB agreement or where one member has signed the agreement and by so doing takes responsibility for the fees of the DAAB."*

The terms of the remuneration of either the sole member or each of the three members, including the remuneration of any expert whom the DAAB consults, shall be mutually agreed by the Parties when agreeing the terms of the DAAB Agreement. Each Party shall be responsible for paying one-half of this remuneration.

This is yet another area where a recalcitrant Party can stymie the Constitution of the DAAB.

The proposal in the previous paragraph is a way of addressing this situation. It is noted that Sub-Clause 21.2 [*Failure to appoint DAAB Members*] does not address this particular issue of failing to agree the level of remuneration.

If at any time the Parties so agree, they may appoint a suitably qualified person or persons to replace any one or more members of the DAAB. Unless the Parties agree otherwise, a replacement DAAB member shall be appointed if a member declines to act or is unable to act as a result of death, illness, disability, resignation or termination of appointment. The replacement member shall be appointed in the same manner as the replaced member was required to have been selected or agreed, as described in this Sub-Clause.

The wording of part of this paragraph is not helpful. A suggested revision is as follows:

> *"If at any time one or more members of the DAAB declines to act or is unable to act as a result of death, disability, resignation or termination of appointment then the Party who appointed that replaced member shall appoint a replacement member in the same manner as the original appointment was made."*

The appointment of any member may be terminated by mutual agreement of both Parties, but not by the Employer or the Contractor acting alone.

This is a useful revision for preventing a unilateral removal of one member of the DAAB.

Unless otherwise agreed by both Parties, the term of the DAAB (including the appointment of each member) shall expire either:

(a) on the date the discharge shall have become, or deemed to have become, effective under Sub-Clause 14.12 [*Discharge*]; or
(b) 28 days after the DAAB has given its decision on all Disputes, referred to it under Sub-Clause 21.4 [*Obtaining DAAB's Decision*] before such discharge has become effective,

whichever is later.

The comma in sub-paragraph (b) should be removed as at present written it qualifies the disputes that do not need qualification.

However, if the Contract is terminated under any Sub-Clause of these Conditions or otherwise, the term of the DAAB (including the appointment of each member) shall expire 28 days after:

(i) the DAAB has given its decision on all Disputes, which were referred to it (under Sub-Clause 21.4 [*Obtaining*

DAAB's Decision]) within 224 days after the date of termination; or

(ii) the date that the Parties reach a final agreement on all matters (including payment) in connection with the termination whichever is earlier.

These Conditions of Contract use brackets in an inconsistent way.

In this paragraph the reference to Sub-Clause 21.4 is placed in a bracket whereas in the previous paragraph there is no bracket. It is therefore recommended that the bracket is removed in this paragraph.

It is also difficult to understand the reference to 224 days in sub-paragraph (i). The DAAB is required to give a decision in 84 days. If there is any concern as to the length of time required for decision, then the sub-paragraph should be silent rather than impose periods of time which is different from that identified in Sub-Clause 21.4 [*Obtaining the DAAB's Decision*].

21.2 Failure to Appoint DAAB Member(s)

If any of the following conditions apply, namely:

(a) if the DAAB is to comprise a sole member, the Parties fail to agree the appointment of this member by the date stated in the first paragraph of Sub-Clause 21.1 [*Constitution of the DAAB*]; or

(b) if the DAAB is to comprise three persons, and if by the date stated in the first paragraph of Sub-Clause 21.1 [*Constitution of the DAAB*]:

(i) either Party fails to select a member (for agreement by the other Party);

(ii) either Party fails to agree a member selected by the other Party; and/or

(iii) the Parties fail to agree the appointment of the third member (to act as chairperson) of the DAAB;

(c) the Parties fail to agree the appointment of a replacement within 42 days after the date on which the sole member or one of the three members declines to act or is unable to act as a result of death, illness, disability, resignation, or termination of appointment; or

(d) if, after the Parties have agreed the appointment of the member(s) or replacement, such appointment cannot be effected because one Party refuses or fails to sign a DAAB Agreement with any such member or replacement (as the case may be) within 14 days of the other Party's request to do so, then the appointing entity or official named in the Contract Data shall, at the request of either or both Parties and after due consultation with both Parties, appoint the member(s) of the DAAB (who, in the case of sub-paragraph (d) above, shall be the agreed member(s) or replacement). This appointment shall be final and conclusive.

Generally, this first paragraph is very helpful. However, as suggested in the analysis of the previous Sub-Clause, this paragraph does not address the situation where the Parties have agreed the appointment or replacement of the member and one Party refuses or failed to sign the appropriate DAAB agreement.

Sub-paragraph (d) refers to a situation where this occurs and does not provide the solution. This sub-paragraph anticipates a situation where the appointing entity or visual is required to appoint the necessary member of DAAB. However, the situation posed in sub-paragraph (d) does not require appointment – the appointment has taken place; what is required is that the appointing entity can deem the DAAB agreement signed on behalf of the recalcitrant Party.

There needs to be a linking provision as follows:

> *"In the case where one Party has refused or failed to sign a DAAB agreement with any appointed or replaced member, as the case may be, then the other Party may request the appointing entity or official named in the Contract Data or in any other relevant document to issue a Notice that the recalcitrant member shall be deemed to have signed and be bound by the DAAB agreement."*

Thereafter, the Parties and the member(s) so appointed shall be deemed to have signed and be bound by a DAAB Agreement under which:

(i) the monthly services fee and daily fee shall be as stated in the terms of the appointment; and

(ii) the law governing the DAAB Agreement shall be the governing law of the Contract defined in Sub-Clause 1.4 [*Law and Language*].

The above paragraph needs to be revised to take account of the situation where an appointment has been made but the agreement has not been endorsed by one Party. The revision is as follows:

> *"Thereafter, the Parties shall be deemed to have signed and be bound by a DAAB agreement under which:*
>
> *(i) the monthly services fee and daily fee shall be as stated in the terms of appointment; and*
> *(ii) the law governing the DAAB agreement shall be the governing law of the Contract defined in Sub-Clause 1.4 [Law and Language]."*

Each Party shall be responsible for paying one-half of the remuneration of the appointing entity or official. If the Contractor pays the remuneration in full, the Contractor shall include one-half of the amount of such remuneration in a Statement and the Employer shall then pay the Contractor in accordance with the Contract. If the Employer pays the remuneration in full, the Engineer shall include one-half of the amount of such remuneration as a deduction under sub-paragraph (b) of Sub-Clause 14.6.1 [*The IPC*].

This is new in that it allows for the situation where the Employer has not paid its share of the fees of the DAAB. It also makes provision for the reverse situation where the Contractor has not paid its share of the fees.

21.3 Avoidance of Disputes

If the Parties so agree, they may jointly request (in writing, with a copy to the Engineer) the DAAB to provide assistance and/or informally discuss and attempt to resolve any issue or disagreement that may have arisen between them during the performance of the Contract. If the DAAB becomes aware of an issue or disagreement, it may invite the Parties to make such a joint request.

Such joint request may be made at any time, except during the period that the Engineer is carrying out his/her duties under Sub-Clause 3.7 [*Agreement or Determination*] on the matter at issue or in disagreement unless the Parties agree otherwise.

> Such informal assistance may take place during any meeting, Site visit or otherwise. However, unless the Parties agree otherwise, both Parties shall be present at such discussions. The Parties are not bound to act on any advice given during such informal meetings, and the DAAB shall not be bound in any future Dispute resolution process or decision by any views or advice given during the informal assistance process, whether provided orally or in writing.

This is a key element of the role of the DAAB although arguably underplayed here in these general Conditions of Contract.

21.4.1 Reference of a Dispute to the DAAB

> The reference of a Dispute to the DAAB (the "reference" in this Sub-Clause 21.4) shall:
>
> (a) if Sub-Clause 3.7 [*Agreement or Determination*] applied to the subject matter of the Dispute, be made within 42 days of giving or receiving (as the case may be) a NOD under Sub-Clause 3.7.5 [*Dissatisfaction with Engineer's determination*]. If the Dispute is not referred to the DAAB within this period of 42 days, such NOD shall be deemed to have lapsed and no longer be valid;
> (b) state that it is given under this Sub-Clause;
> (c) set out the referring Party's case relating to the Dispute;
> (d) be in writing, with copies to the other Party and the Engineer; and
> (e) for a DAAB of three persons, be deemed to have been received by the DAAB on the date it is received by the chairperson of the DAAB.
>
> The reference of a Dispute to the DAAB under this Sub-Clause shall, unless prohibited by law, be deemed to interrupt the running of any applicable statute of limitation or prescription period.

This Sub-Clause is new. Sub-paragraph (a) does not identify any disputes that do not fall under Sub-Clause 3.7 [*Agreement or Determination*]. It should add:

> *Where the dispute does not fall under Sub-Clause 3.7 [Agreement or Determination] then a reference may be made at any time.*

21.4.2 The Parties' obligations after the reference

> Both Parties shall promptly make available to the DAAB all information, access to the Site, and appropriate facilities, as the DAAB may require for the purposes of making a decision on the Dispute.
>
> Unless the Contract has already been abandoned or terminated, the Parties shall continue to perform their obligations in accordance with the Contract.

This is a useful addition to these Conditions of Contract.

21.4.3 The DAAB's decision

> The DAAB shall complete and give its decision within:
>
> (a) 84 days after receiving the reference; or
> (b) such period as may be proposed by the DAAB and agreed by both Parties.
>
> However, if at the end of this period, the due date(s) for payment of any DAAB member's invoice(s) has passed but such invoice(s) remains unpaid, the DAAB shall not be obliged to give its decision until such outstanding invoice(s) have been paid in full, in which case the DAAB shall give its decision as soon as practicable after payment has been received.

This has been inserted, as one of the major practical problems for a DAAB is the unwillingness of one of the Parties to pay the outstanding invoices. By inserting this paragraph in the general Conditions of Contract there is a mechanism to ensure that all members of the DAAB have received payment for the work that they have done and that the Parties cannot benefit from that work until they have met their obligations.

The decision shall be given in writing to both Parties with a copy to the Engineer, shall be reasoned, and shall state that it is given under this Sub-Clause.

The reference to writing in this context usually means a written document but can of course mean that electronic communication can be equally acceptable. There is no definition of a reasoned award or decision. It is believed that the general instruction to DAABs preparing a reasoned decision is as follows:

set out the names of the Parties,
the background scenario,
the dispute,
the evidence of the Claimant followed by evidence of the other Party, further information sought by the DAAB,
the Site visit,
any expert evidence which the DAAB may require,
and finally the decision itself.

The decision itself must make clear not only the history of the Dispute or disagreement but also allocates blame or fault so that each Party knows from the DAAB considers responsible.

The decision shall be binding on both Parties, who shall promptly comply with it whether or not a Party gives a NOD with respect to such decision under this Sub-Clause. The Employer shall be responsible for the Engineer's compliance with the DAAB decision.

This important paragraph is not absolutely clear. It should stress that all Parties should comply with the decision. The amendment should be as follows:

> *"The decision shall be binding on all Parties and the Engineer, all of whom shall promptly comply with it whether or not Party gives an end with respect to such a decision under the Sub-Clause."*

If the decision of the DAAB requires a payment of an amount by one Party to the other Party

(i) subject to sub-paragraph (ii) below, this amount shall be immediately due and payable without any certification or Notice; and

(ii) the DAAB may (as part of the decision), at the request of a Party but only if there are reasonable grounds for the DAAB to believe that the payee will be unable to repay such amount in the event that the decision is reversed under Sub-Clause 21.6 [*Arbitration*], require the payee to provide an appropriate security (at the DAAB's sole discretion) in respect of such amount.

This last addition to the powers of the DAAB is very important. The Clause does not make clear when the Party should make the application to the DAAB for security in advance of the receipt of the payment. A Party effectively advised would make such a request during the DAAB proceedings for security in advance of the receipt of the payment. However, the fact that the paragraph is silent on the matter could imply that a request for security could be made after the decision has been issued.

An Amendment is proposed to address this gap in the procedure as follows:

> *"In the event that a Party anticipate that it will require the DAAB to order security for the payment will be required if it is appropriate for the Party seeking such a Security to make this application before the DAAB issues its decision. However, if the Party fails to make such an application until after the decision has been issued the Party may make a late application but must be responsible for the costs of the DAAB in providing an additional decision."*

The DAAB proceeding shall not be deemed to be an arbitration and the DAAB shall not act as arbitrator(s).

This paragraph on its own is especially meaningless and not necessarily correct. The DAAB may ask questions of their own initiative, take evidence, call witnesses, and of course write a reasoned award or decision. It is also rather frequent that the DAAB is requested by the Agreement to act as an arbitration court up to a certain limit amount.

21.4.4 Dissatisfaction with DAAB's decision

If either Party is dissatisfied with the DAAB's decision:

(a) such Party may give a NOD to the other Party, with a copy to the DAAB and to the Engineer;

(b) this NOD shall state that it is a "Notice of Dissatisfaction with the DAAB's Decision" and shall set out the matter in Dispute and the reason(s) for dissatisfaction; and

(c) this NOD shall be given within 28 days after receiving the DAAB's decision.

This is a clear and established procedure. It requires careful consideration by the Party who is dissatisfied. The challenge must be effective and well-reasoned. And of great importance is the 28-day period. If that 28-day period has passed then basically the Party who is dissatisfied is effectively disbarred from challenging the DAAB's decision.

It is also vital to remember that the decision may include various elements and the Notice of Dissatisfaction may only apply to one of two of those elements. The fact that it is only a challenge to part of the Decision should not and must not prevent the dissatisfied Party from issuing the Notice of Dissatisfaction.

If the DAAB fails to give its decision within the period stated in Sub-Clause 21.4.3 [*The DAAB's decision*], then either Party may, within 28 days after this period has expired, give a NOD to the other Party in accordance with sub-paragraphs (a) and (b) above.

It seems the more paradoxical to be able to challenge a decision that does not exist. However, it is an excellent procedure because it ensures that either the DAAB gives its decision within the appropriate time period or, if it fails to do so, then the dissatisfied Party can immediately, after the appropriate time has elapsed, require the matter to be determined before the Arbitration tribunal. In fact, the heading to 21.4.4 requires an amendment as follows:

"Dissatisfaction with the DAAB's decision or absence of decision"

Except as stated in the last paragraph of Sub-Clause 3.7.5 [*Dissatisfaction with Engineer's determination*], in Sub-Clause 21.7 [*Failure to Comply with DAAB's Decision*] and in Sub-Clause 21.8 [*No DAAB In Place*], neither Party shall be entitled to commence arbitration of a Dispute unless a NOD in respect of that Dispute has been given in accordance with this Sub-Clause 21.4.4.

This is a very useful procedural step described so as to ensure that once a Party has issued a Notice of Dissatisfaction it has 182 days, too long in some opinions, to refer the matter or dispute to arbitration.

If the DAAB has given its decision as to a matter in Dispute to both Parties, and no NOD under this Sub-Clause 21.4.4 has been given by either Party within 28 days after receiving the DAAB's decision, then the decision shall become final and binding on both Parties.

The final comment as to final and binding is put in the present tense not in the future tense. The future tense implies that the procedure requires another action. In fact, it does not. Therefore, it is suggested that this paragraph is amended as follows:

> *"Where the DAAB has given its decision as to a matter in dispute to both Parties, and no NOD under this Sub-Clause 21.4.4 has been given by either Party within 28 days after receiving a DAAB's decision then the decision is final and binding on both Parties."*

If the dissatisfied Party is dissatisfied with only part(s) of the DAAB's decision:

(i) this part(s) shall be clearly identified in the NOD;
(ii) this part(s), and any other parts of the decision that are affected by such part(s) or rely on such part(s) for completeness, shall be deemed to be severable from the remainder of the decision; and
(iii) the remainder of the decision shall become final and binding on both Parties as if the NOD had not been given.

This is a very useful addition to the Conditions.

21.5 Amicable Settlement

Where a NOD has been given under Sub-Clause 21.4 [*Obtaining DAAB's Decision*], both Parties shall attempt to settle the Dispute amicably before the commencement of arbitration. However, unless both Parties agree otherwise, arbitration may be commenced on or after the twenty-eighth (28th) day after the day on which this NOD was given, even if no attempt at amicable settlement has been made.

This is a very useful direction to the Parties except it puts the onus on the Parties and does not involve the DAAB. Often Parties at this stage of a dispute have no wish to communicate with each other. It will be recorded that the DAAB has an opportunity to involve itself informally to attempt to resolve matters or disagreements as an early stage.

This paragraph should be amended as follows:

> *"Where a NOD has been given under Sub-Clause 21.4 [Obtaining DAAB's Decision], both Parties shall attempt to settle the Dispute amicably before the commencement of Arbitration, such Amicable Settlement may be limited to the Parties or may permit the Parties to request the assistance of the DAAB to act in its informal capacity. However, unless both Parties agree otherwise, Arbitration may be commenced on or after the twenty eighth (28th) day after the date on which this NOD was given, even if no attempt at Amicable Settlement has been made."*

21.6 Arbitration

Unless settled amicably, and subject to Sub-Clause 3.7.5 [*Dissatisfaction with Engineer's determination*], Sub-Clause 21.4.4 [*Dissatisfaction with DAAB's decision*], Sub-Clause 21.7 [*Failure to Comply with DAAB's Decision*] and Sub-Clause 21.8 [*No DAAB In Place*], any Dispute in respect of which the DAAB's decision (if any) has not become final and binding shall be finally settled by international arbitration. Unless otherwise agreed by both Parties:

(a) the Dispute shall be finally settled under the Rules of Arbitration of the International Chamber of Commerce;
(b) the Dispute shall be settled by one or three arbitrators appointed in accordance with these Rules; and
(c) the arbitration shall be conducted in the ruling language defined in Sub-Clause 1.4 [*Law and Language*].

It should be noted that some experienced Parties are opting for arbitration other than under the auspices of the International Chamber of Commerce. Such decisions should be made part of the Contract Data or Employer's Requirements. The second sub-paragraph (b) states that the Dispute shall be settled by one or three

Arbitrators appointed in accordance with the ICC rules. It is recommended that this sub-paragraph be amended as follows:

> *"The dispute shall be settled by three arbitrators, unless the Parties agree to the contrary, appointed in accordance with these rules."*

Once the matter has reached the decision-making process of Arbitration then it is preferable to have the wider experience available provided by a three-man arbitration tribunal than a single Arbitrator.

The arbitrator(s) shall have full power to open up, review and revise any certificate, determination (other than a final and binding determination), instruction, opinion or valuation of the Engineer, and any decision of the DAAB (other than a final and binding decision) relevant to the Dispute. Nothing shall disqualify the Engineer from being called as a witness and giving evidence before the arbitrator(s) on any matter whatsoever relevant to the Dispute.

It is important to note that the same power to call the Engineer as a witness does not extend to the members of the DAAB. At least that appears to be the case. The General Conditions of Contract are silent on the issue. There should be an amendment as follows:

> *"Where the Arbitration tribunal consider it necessary for the effective Decision-making process it is empowered to call members of the DAAB to give evidence before the Arbitrator(s) on any matter whatsoever relevant to the Dispute."*

In any award dealing with costs of the arbitration, the arbitrator(s) may take account of the extent (if any) to which a Party failed to cooperate with the other Party in constituting a DAAB under Sub-Clause 21.1 [*Constitution of the DAAB*] and/or Sub-Clause 21.2 [*Failure to Appoint DAAB Member(s)*].

The sanctions available should be increased to include the situation where a Party has failed to sign the DAAB agreement under Sub-Clause 21.2 Sub-paragraph (d).

The amendment proposed is as follows:

> *"In any award dealing with costs of the arbitration, the arbitrators may take account of the extent (if any) to which a Party failed to cooperate with the other Party in constituting a DAAB under Sub-Clause 21.1 [Constitution of the DAAB], Sub-Clause 21.2 [Failure to appoint DAAB Members] and/ or Sub-Clause 21.2 [Failure to sign the DAAB Agreement]."*

Neither Party shall be limited in the proceedings before the arbitrator(s) to the evidence or arguments previously put before the DAAB to obtain its decision, or to the reasons for dissatisfaction given in the Party's NOD under Sub-Clause 21.4 [*Obtaining DAAB's Decision*]. Any decision of the DAAB shall be admissible in evidence in the arbitration.

This is a key paragraph in this Sub-Clause. It provides a Party with an opportunity to retrospectively examine the project and, with the benefit of hindsight, identify the crucial areas, which, if dealt with effectively, would have prevented disputes arising. Some arbitral tribunals attempt to identify one issue in spite of the fact that a Party clearly has sought a review of the entire series of disputes throughout the project.

Arbitration may be commenced before or after completion of the Works. The obligations of the Parties, the Engineer and the DAAB shall not be altered by reason of any arbitration being conducted during the progress of the Works.

In spite of the direction to all participants in the project it is inevitable that there will be pressures upon the Parties as a result of the arbitral process. It is up to the Engineer to attempt to manage the relationships in a proactive way to ensure that the progress of the execution of the Works is not hampered by any Arbitration.

If an award requires a payment of an amount by one Party to the other Party, this amount shall be immediately due and payable without any further certification or Notice.

This is a crucial variation to the general Conditions of Contract. Once the arbitral Award has defined the issues and where appropriate ordered a payment to be made, that payment can be invoiced immediately.

21.7 Failure to Comply with DAAB's Decision

In the event that a Party fails to comply with any decision of the DAAB, whether binding or final and binding, then the other Party may, without prejudice to any other rights it may have, refer the failure itself directly to arbitration under Sub-Clause 21.6 [*Arbitration*] in which case Sub-Clause 21.4 [*Obtaining DAAB's Decision*] and Sub-Clause 21.5 [*Amicable Settlement*] shall not apply to this reference. The arbitral tribunal (constituted under Sub-Clause 21.6 [*Arbitration*]) shall have the power, by way of summary or other expedited procedure, to order, whether by an interim or provisional measure or an award (as may be appropriate under applicable law or otherwise), the enforcement of that decision.

The references to binding and final and binding without reference to the particular Sub-Clause could be confusing. The first sentence needs breaking into two sentences in order to avoid confusion. There is a useful direction to the arbitral tribunal that the power it may have under the ICC rules may be amplified by use of a summary or expedited procedure. However, the essence of the Arbitration Clause in the general Conditions of Contract is that the tribunal may open up and review the entire process of Dispute Resolution.

By authorising the tribunal to utilise a summary or expedited procedure questions could be raised as to the effectiveness of an Arbitral tribunal which has already ordered the enforcement of a DAAB Decision. Can the tribunal retain its required impartiality after implementing summary enforcement of a DAAB Decision now being challenged?

The amended first paragraph of this Sub-Clause is as follows:

> *"In the event that a Party has failed to comply with any decision of the DAAB, whether binding under Sub-Clause 21.4.3 [The DAAB's Decision] or final and binding under Sub-Clause 21.4.4 [Dissatisfaction with DAAB's Decision],*
>
> *then the other Party may, without prejudice to any rights it may have, refer the Decision, which has not been complied with, directly to Arbitration under Sub-Clause 21.6 [Arbitration]. Sub-Clause 21.4 [Obtaining the DAAB's Decision] and Sub-Clause 21.5 [Amicable Settlement] shall not apply to this reference. The arbitral tribunal (constituted under Sub-Clause 21.6 [Arbitration]) shall have the power, either by way of summary or expedited procedure, to order, whether by way of an interim award or other enforceable measure (as may be appropriate under the applicable Law or otherwise), the enforcement of that Decision."*

In the case of a binding but not final decision of the DAAB, such interim or provisional measure or award shall be subject to the express reservation that the rights of the Parties as to the merits of the Dispute are reserved until they are resolved by an award.

> Any interim or provisional measure or award enforcing a decision of the DAAB, which has not been complied with, whether such decision is binding or final and binding, may also include an order or award of damages or other relief.

The paragraph has avoided mentioning the status of the decision of the DAAB, which is final and binding. It is clear from the Arbitration Clause 21.6 [*Arbitration*] that any ruling or decision of the DAAB relevant to this dispute can be opened up, reviewed and revised. These General Conditions of Contract have not excluded the authority of the arbitral tribunal to review a decision of the DAAB, which is final and binding. These General Conditions of Contract need to make such a provision if the proposals are to be effective. However, it is strongly recommended that the wide powers given to the arbitral tribunal are not restricted in the manner implied by this paragraph.

There is a revision proposed to clarify the award referred to in the last line of this paragraph. The revision is as follows:

> *"In the case of a binding but not final decision of the DAAB, such interim or provisional measure or award shall be subject to the express reservation that the rights of the Parties as to the merits of the dispute are reserved until they are resolved by an award determinative of all the issues before the arbitral tribunal."*

It is emphasised that silence as to the status of the final and binding DAAB Decision does not of itself ensure that such a Decision cannot be reviewed by an Arbitral tribunal.

Any interim or provisional measure or Award enforcing a Decision of the DAAB, which has not been complied with, whether such a Decision is binding or final and binding, may also order more damages or other relief.

This paragraph, which it is believed correctly sets out the position as to the review of a final and binding Decision of the DAAB, referred to above, is important in that it expands the powers of the arbitral tribunal.

21.8 No DAAB In Place

> If a Dispute arises between the Parties in connection with, or arising out of, the Contract or the execution of the Works and there is no DAAB in place (or no DAAB is being constituted), whether by reason of the expiry of the DAAB's appointment or otherwise:
>
> (a) Sub-Clause 21.4 [*Obtaining DAAB's Decision*] and Sub-Clause 21.5 [*Amicable Settlement*] shall not apply; and

(b) the Dispute may be referred by either Party directly to arbitration under Sub-Clause 21.6 [*Arbitration*] without prejudice to any other rights the Party may have.

The reference to there being no DAAB in place is confusing. What is intended is that where the process for establishing the DAAB has not been fully complied with then the Parties can refer any dispute directly to Arbitration. The DAAB may well be in place but the agreements may not have been signed by all Parties.

This paragraph needs to be amended as follows:

"If a dispute arises between the Parties in connection with, or arising out of, the Contract or the execution of the Works and the procedure for establishing the DAAB has not been fully complied with then:

(a) Sub-Clause 21.4 [Obtaining DAAB's Decision] and Sub-Clause 21.5 [Amicable Settlement] shall not apply; and
(b) the dispute may be referred by either Party directly to arbitration under Sub-Clause 21.6 [Arbitration] without prejudice to any other rights he may have."

INDEX